D1228813

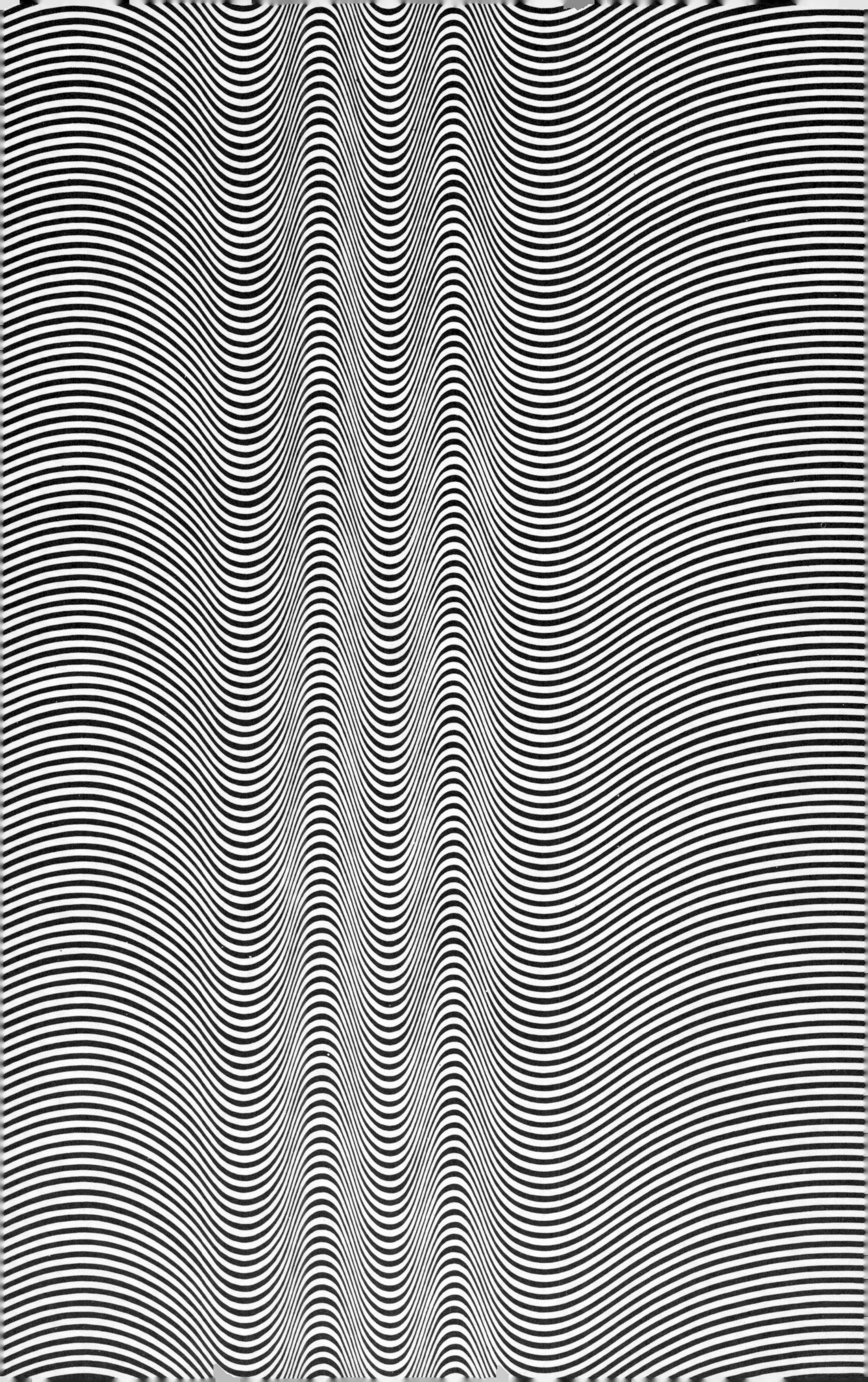

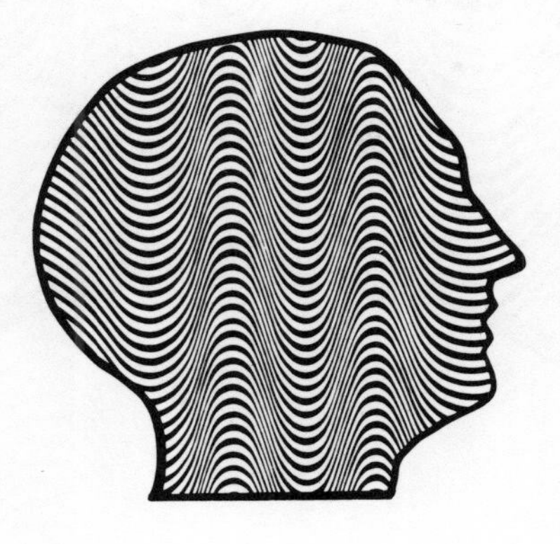

Biorhythm

#4651

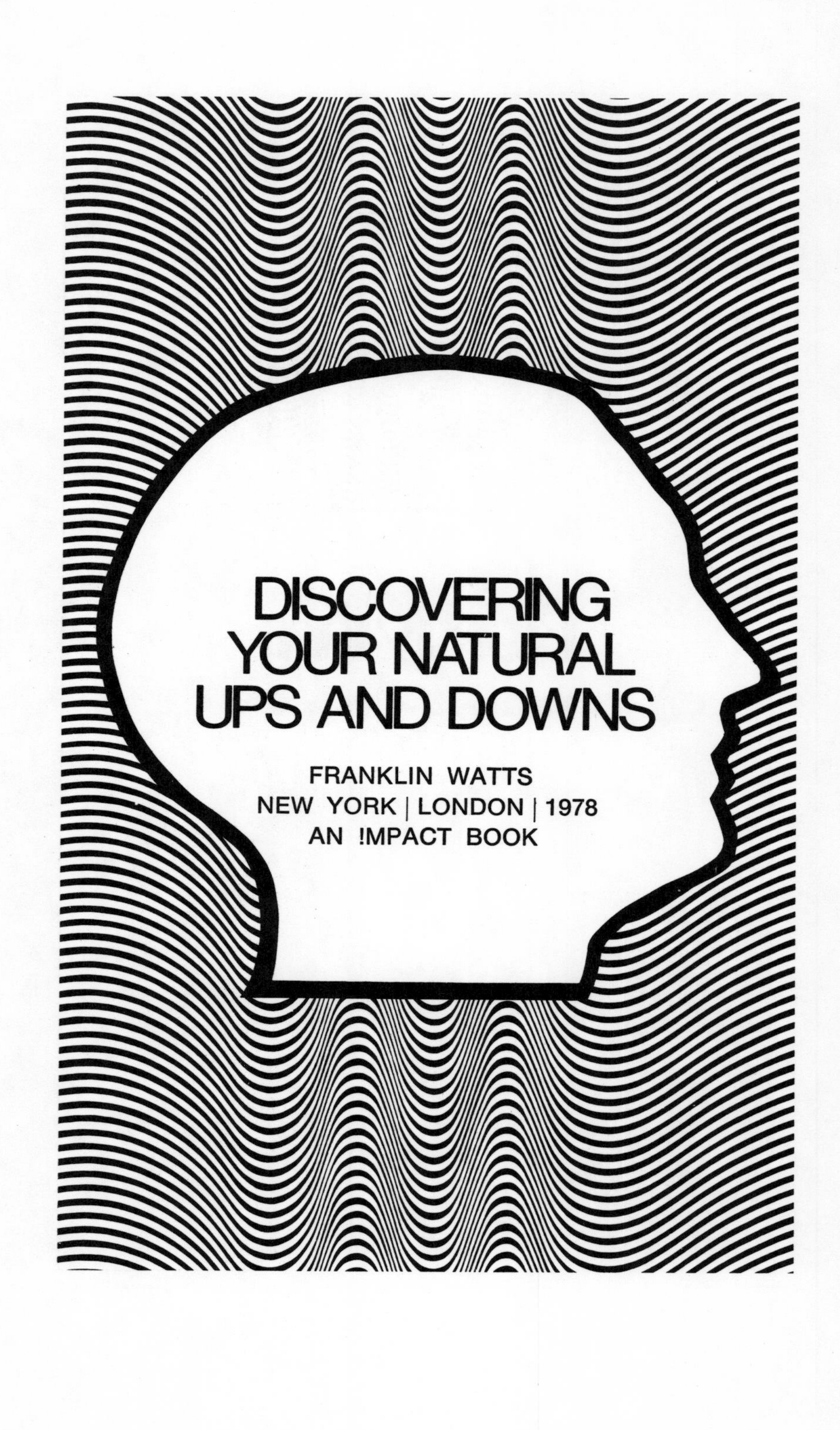
DISCOVERING
YOUR NATURAL
UPS AND DOWNS
FRANKLIN WATTS
NEW YORK | LONDON | 1978
AN !MPACT BOOK

Biorhythm
Biorhythm
Biorhythm

BY PAULINE C. BARTEL

Biorhythm
Biorhythm
Biorhythm

Diagrams by Vantage Art, Inc.

Library of Congress Cataloging in Publication Data

Bartel, Pauline C

Biorhythm: discovering your natural ups and downs.

(An Impact book)
Bibliography: p.
Includes index.
SUMMARY: Explains the theory of biorhythm and its relationship to rhythms in nature and provides instructions for calculating one's own rhythm positions.
1. Biological rhythms—Juvenile literature. [1. Biological rhythms] I. Title.
BF637.B55B37 001.9 77-17585
ISBN 0-531-01355-3

Printed in the United States of America
5 4 3 2 1

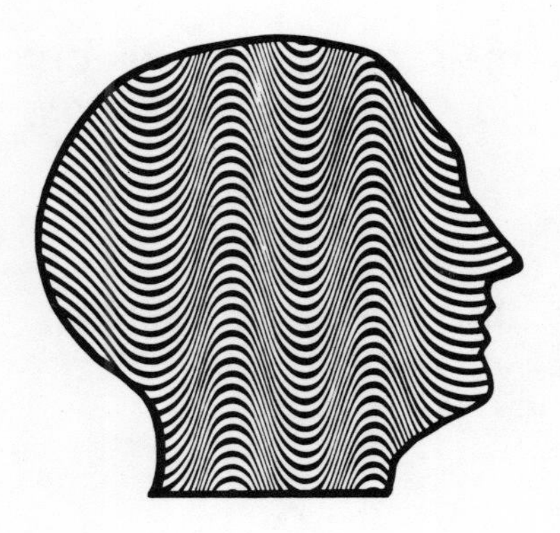

CONTENTS

This book is dedicated to the three forces that,
like the three biorhythms, have directed my life:

To my parents,
for their support and encouragement
To my husband, Hugh McAteer,
for his patience and understanding
To my mentor, Arnold Madison,
for his knowledge and guidance

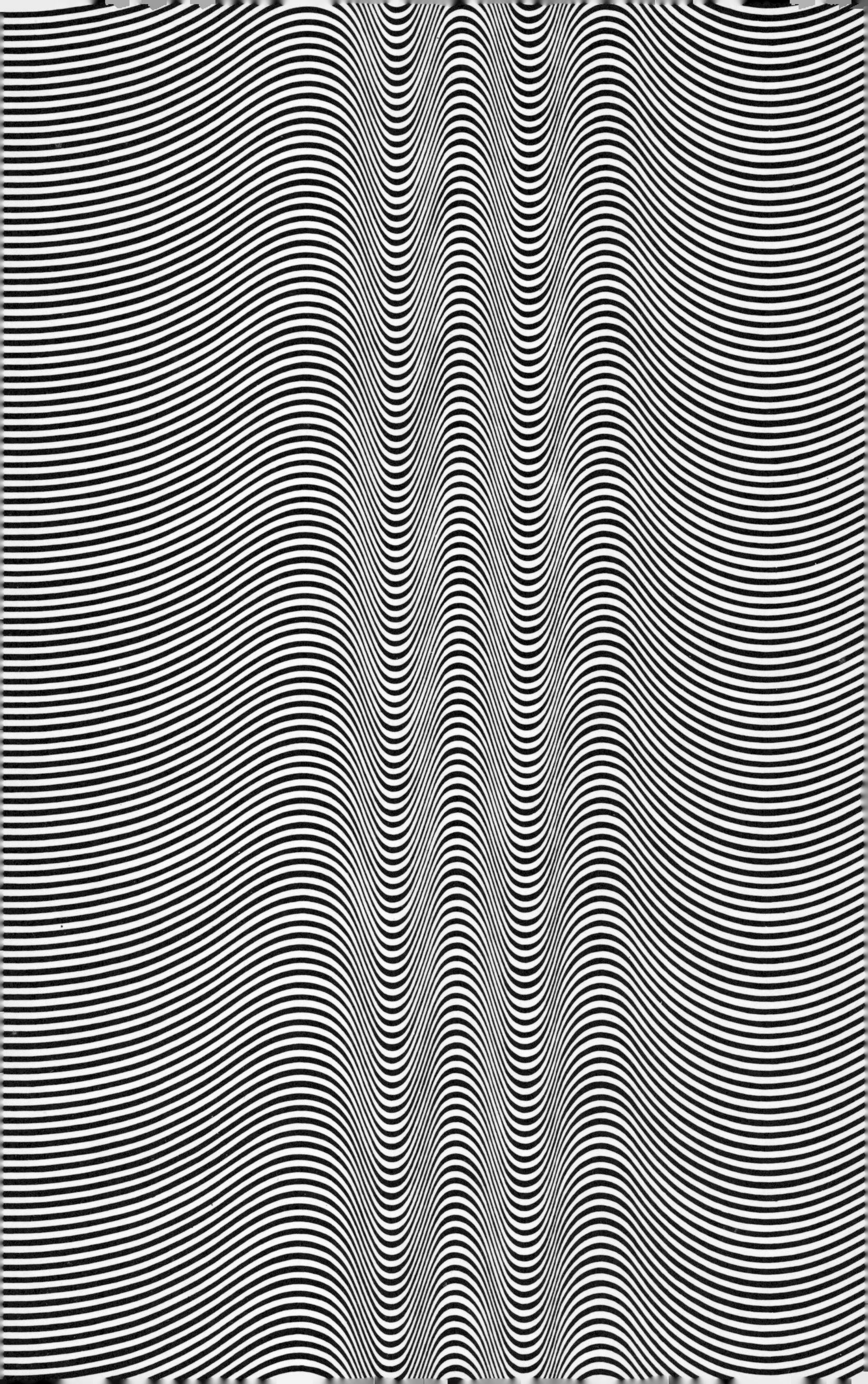

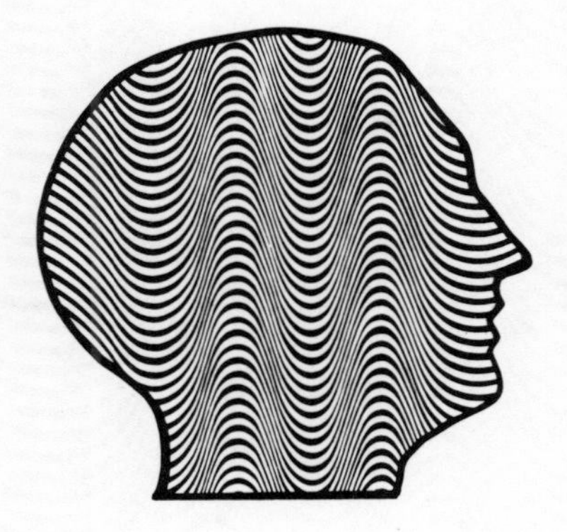

Biorhythm

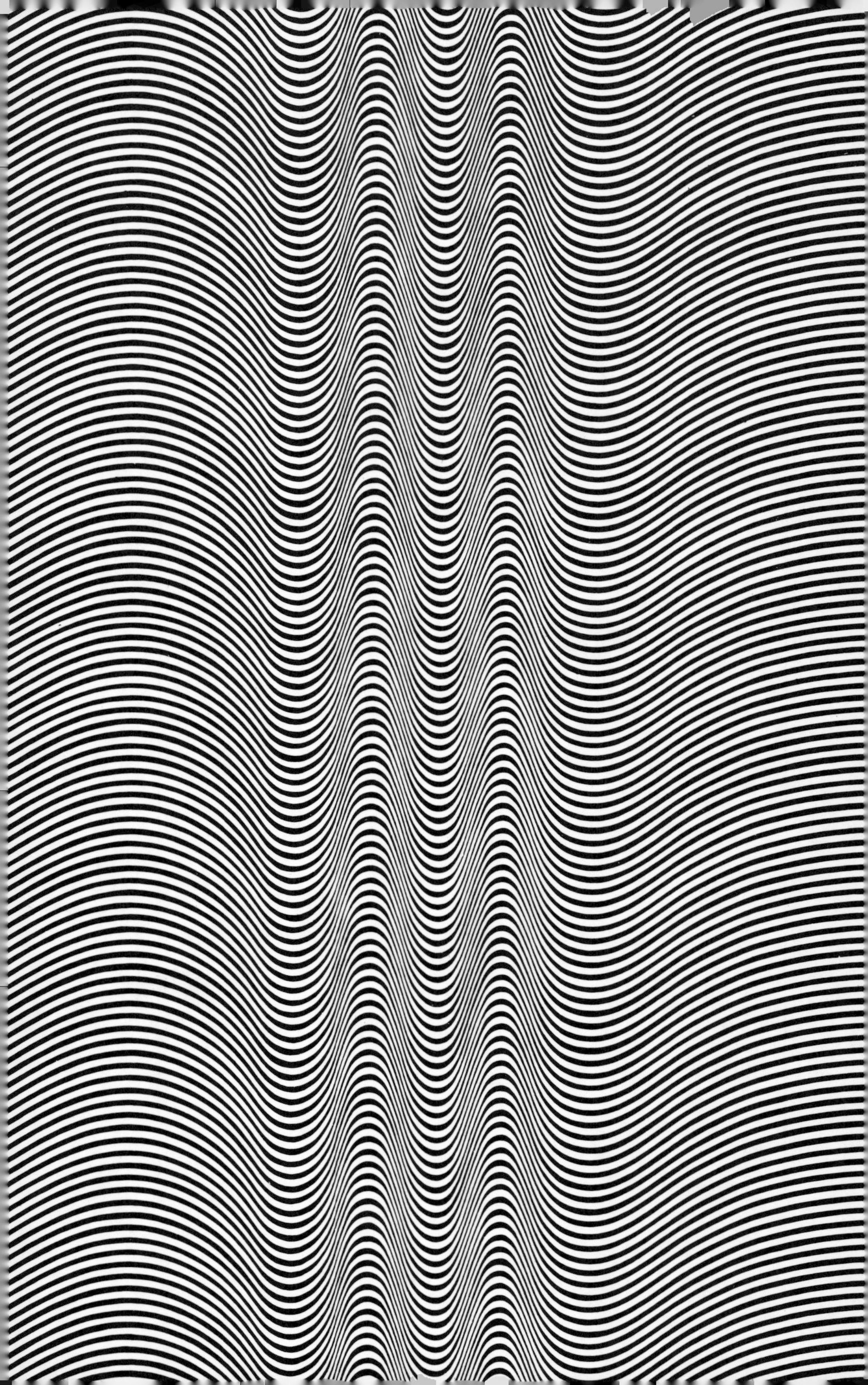

BIORHYTHM: FROM THE MOMENT OF BIRTH

1

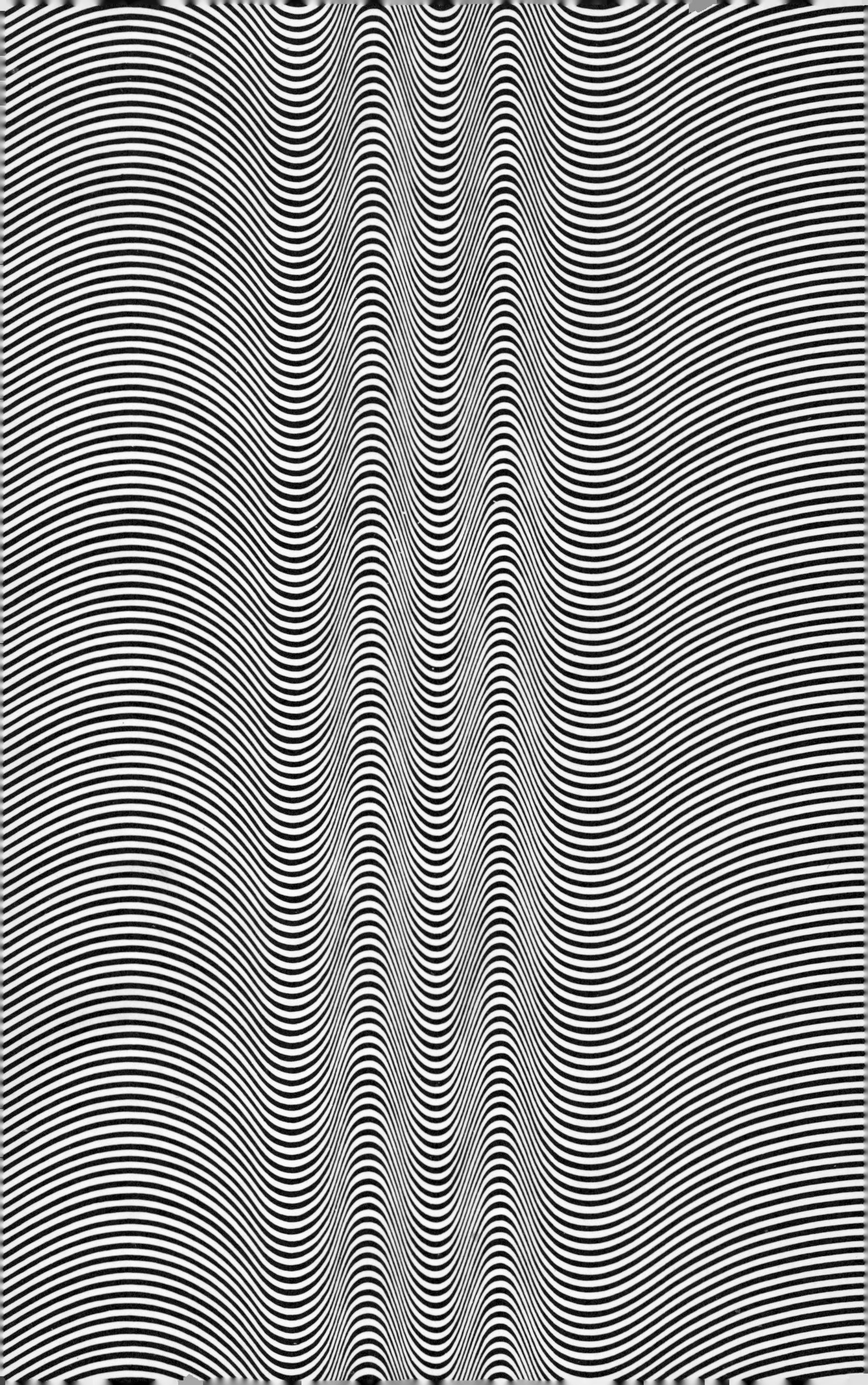

"What a terrible day I had today. I would have been better off staying in bed."

So begins the lament that all of us have experienced at one time or another. There is a way, however, to know in advance which days of the month may be "good" days and which may be "bad" days. The method is through the theory of life cycles known as *biorhythm.*

The biorhythm theory states that, from the moment of birth, each person is influenced by three separate rhythms: a 23-day *physical rhythm;* a 28-day *emotional,* or *sensitivity, rhythm;* and a 33-day *intellectual rhythm.* In each of the three rhythms, half of the days are *plus* days in which the individual is at his or her best. The remaining half of the days are *minus* days. During that portion of the cycle, efficiency is reduced. In effect, the body recuperates.

In all three rhythms, *critical* days occur on day 1 of a new cycle and when the rhythm goes from plus to minus at mid-cycle. These are days of instability when the cycle is neither up or down. More care should be taken during critical days, since biorhythm studies have shown that most accidents occur and illnesses begin on these days, which comprise about 20 percent of one's life. (The remaining 80 percent are days of mixed rhythms.) Critical days are not to be feared in themselves. They are merely days in which one's reaction to the world may make one more vulnerable to critical situations. A more thorough understanding of the biorhythm theory may be achieved through examination of each of the three rhythms.

The 23-day physical rhythm affects energy, endurance, resistance to disease, and strength. The 11½ plus days are conducive to any activity that requires peak physical performance. This is an excellent time for intensive athletic training. During the 11½ minus days, a person may tire more quickly and feel less energetic. It would be better for athletes to practice routine exercises than to risk overtraining during this period. Critical days in this rhythm may be accident-prone days. More care should be taken in performing daily activities on a critical day.

The emotional cycle of 28 days affects feelings and creativity. During the 14 plus days, a cheerful outlook on life exists. When two rhythms are in the plus phase simultaneously, remarkable accomplishments are possible. Mark Spitz was under the influence of such a high period in both his physical and emotional rhythms when he won seven gold medals in the 1972 Olympic Games. But the 14 minus days of the emotional rhythm often find a person irritable and negative. Critical days during this rhythm should also be approached cautiously. Emotional outbursts toward parents or friends, for example, are likely. The ability to react to situations quickly and with good judgment is also impaired. Drivers should be especially careful on critical days.

The interesting aspect of this rhythm is that the critical days will always be on the weekday of one's birth. So someone born on a Wednesday can anticipate a critical day every other Wednesday. By being aware of this fact, such a person can avoid a Woeful Wednesday by taking extra care on the first and fifteenth days of this rhythm.

When a critical day in this rhythm coincides with a critical day in the physical rhythm, the day is referred to as a *double critical day,* and the possibility of error or accident is increased. Double critical days occur about six times a year.

Mental alertness, reasoning power, and logic are affected by the 33-day intellectual rhythm. During the 16½ plus days, the individual performs intellectual activities better, since memory functions at its peak, and mental response is keen. This is an excellent time in which to study new subjects or tackle mathematics. Learning is easier, and the retention of learned material is increased during the plus days. The minus days are good days for practice or review because intellectual capacities are reduced. During critical days, it would be better to postpone making important decisions. The critical days in this rhythm are not considered as important as those in the physical or emotional rhythm unless the day coincides with the critical day of one of the other two rhythms. A *triple critical day* occurs when all

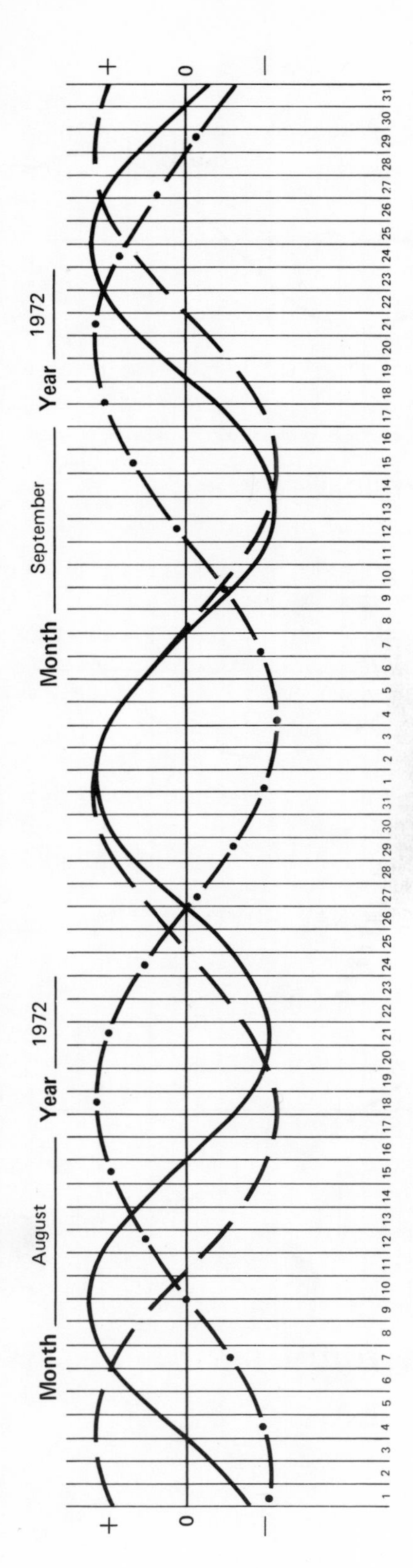

Mark Spitz was under the influence of a double high period when he won seven gold medals at the 1972 Olympic Games. His date of birth is February 10, 1950. The gold medals were won between August 28 and September 4.

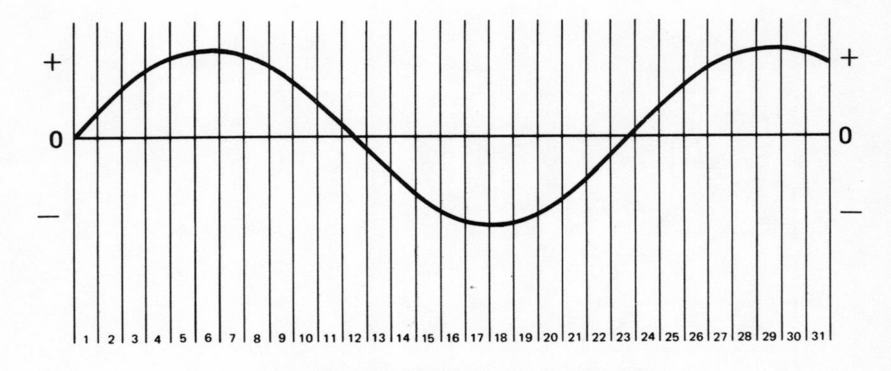

THE 23-DAY PHYSICAL RHYTHM

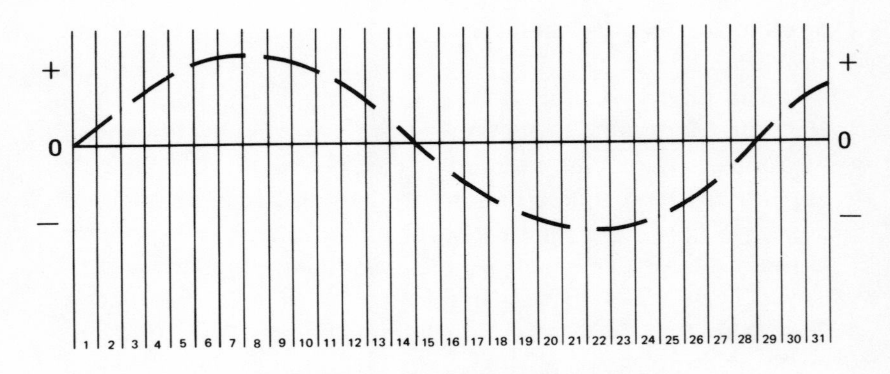

THE 28-DAY EMOTIONAL RHYTHM

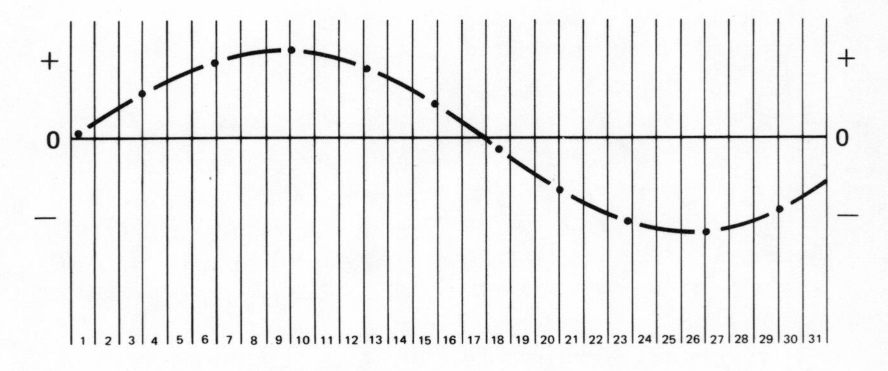

THE 33-DAY INTELLECTUAL RHYTHM

three rhythms are beginning new cycles or switching into minus on the same day. This happens only about once a year.

Critical days can also be potentially dangerous when the cycles are in conflict. An emotional or intellectual critical day in combination with a physical high could produce an accident. In that case, the person might not be alert or use common sense at a time when he or she is exceptionally strong or energetic.

When the biorhythm theory is used, the individual's personality, age, state of health, and temperament must be considered. There are people who are generally more accident-prone and others who are easily irritated. The interpretation of biorhythm charts, therefore, is strictly on an individual basis.

The theory of biorhythm is not new. Between 1897 and 1902 Dr. Hermann Swoboda, a professor of psychology at the University of Vienna, studied the possibility of calculating the rhythmical fluctuations of human feelings and actions. He noted the recurrence of pain, as well as a rhythm in heart attacks, illnesses, and fevers. His findings were published in 1904. He stated the existence of a 23-day and a 28-day rhythm, both affecting human beings from the moment of birth. He also designed a slide rule which, using a person's birth date, could detect the critical days in that person's life.

Amazingly, research into the 23-day and 28-day rhythms was being carried out simultaneously and independently in Berlin by a nose and throat specialist, Dr. Wilhelm Fliess. He gathered research findings to confirm the existence of the two rhythms, which he had observed while treating his patients. Dr. Fliess determined that the 23-day rhythm, originating in muscular cells, affected the physical side of human beings. Having its origins in the nervous system, the 28-day rhythm, he concluded, influenced the emotions or sensitivity of human beings.

It was not until the 1920s that the third rhythm, that of the mind or intelligence, was discovered. Alfred Teltscher was a doctor of engineering and a teacher at Innsbruck, Austria. He tried to find the reason for the fluctuation in the intellectual achievements of students. There is no original record of his research except for articles that discuss his findings. But he discovered that a precise 33-day cycle expressed the high and low points of student performance.

Similar research was conducted at the University of Pennsylvania between 1928 and 1932. While recording the reactions of workers in railroad shops over a length of time, Dr. Rexford Hersey and his assistant, Dr. Michael John Bennett, made the discovery of a 33-day rhythm.

The origin of the intellectual rhythm is in dispute. The two schools of thought are that the rhythm either begins in the cells of the brain or results from a thyroid gland secretion.

It is encouraging that the concurrent, independent research in the field of biorhythm has yielded such consistent data. It strengthens the case that biorhythm theory can help us understand—and control—our lives.

An interesting aspect has been the application of biorhythm to the study of accidents. In 1939 Hans Schwing was a student at the Swiss Federal Institute of Technology in Zurich. Schwing was interested in the biorhythm approach to accident research. He obtained details of 700 accident cases from Swiss insurance companies and applied biorhythm calculations to them. The results were startling. Of the 700 cases, he found that 401 of the accidents occurred on critical days. Therefore, of those accidents, nearly 60 percent happened on critical days (which make up only about 20 percent of one's life).

Biorhythm does not predict accidents. Biorhythm calculations are merely guides for the individual to the days when he or she should be more aware of his or her actions. Take, for example, the everyday action of walking. Few of us stop to think about the procedure of putting one foot in front of the other in a continuous manner until we reach our destination. The act of walking has become automatic, just as answering the telephone or opening a door are automatic responses. If we had to think of every single action, there would be little time to think of anything else. Most of our actions are so automatic that, when the instability of a critical day disturbs these acts, there is a greater likelihood of accidents due to human error.

Biorhythm is clearly a way to anticipate our ups and downs. But if a cycle is down or a critical day is on the horizon, there is no need for one to draw the covers over one's head until a better day dawns. One can still face the "not so terrific" day by being prepared in advance to be more careful or work harder. That monster history exam won't seem so ferocious on a critical day if one gets extra studying in or grabs additional hours of sleep the night before. Biorhythm can help us avoid error and accidents by making us more aware of our "bad" days and can help us take advantage of our "good" days.

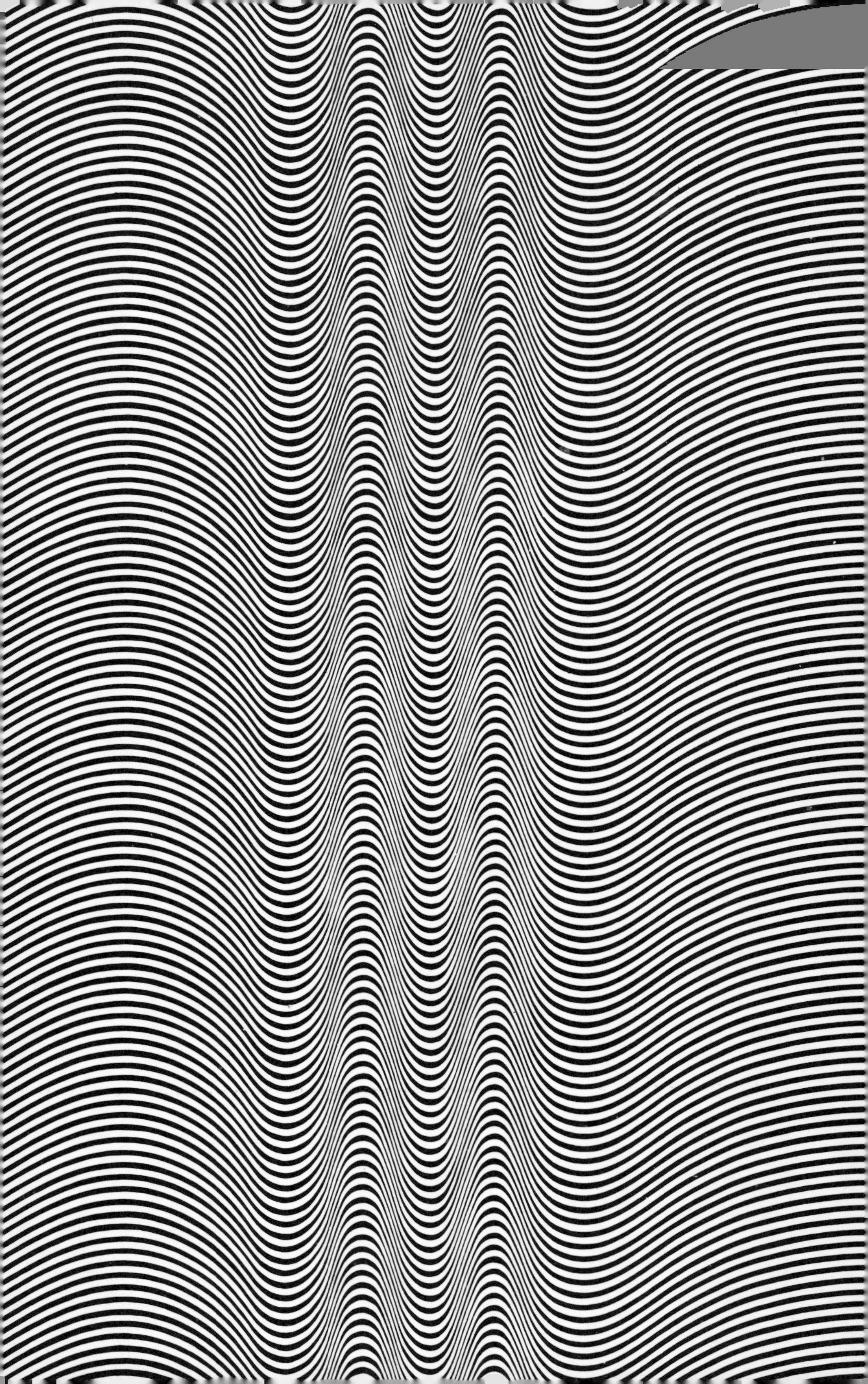

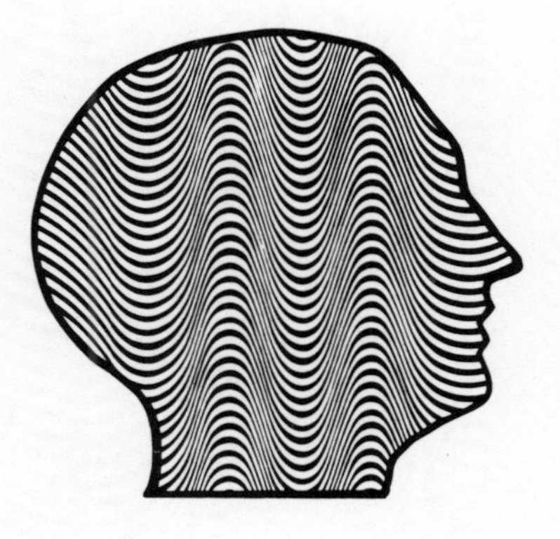

RHYTHMS IN NATURE

2

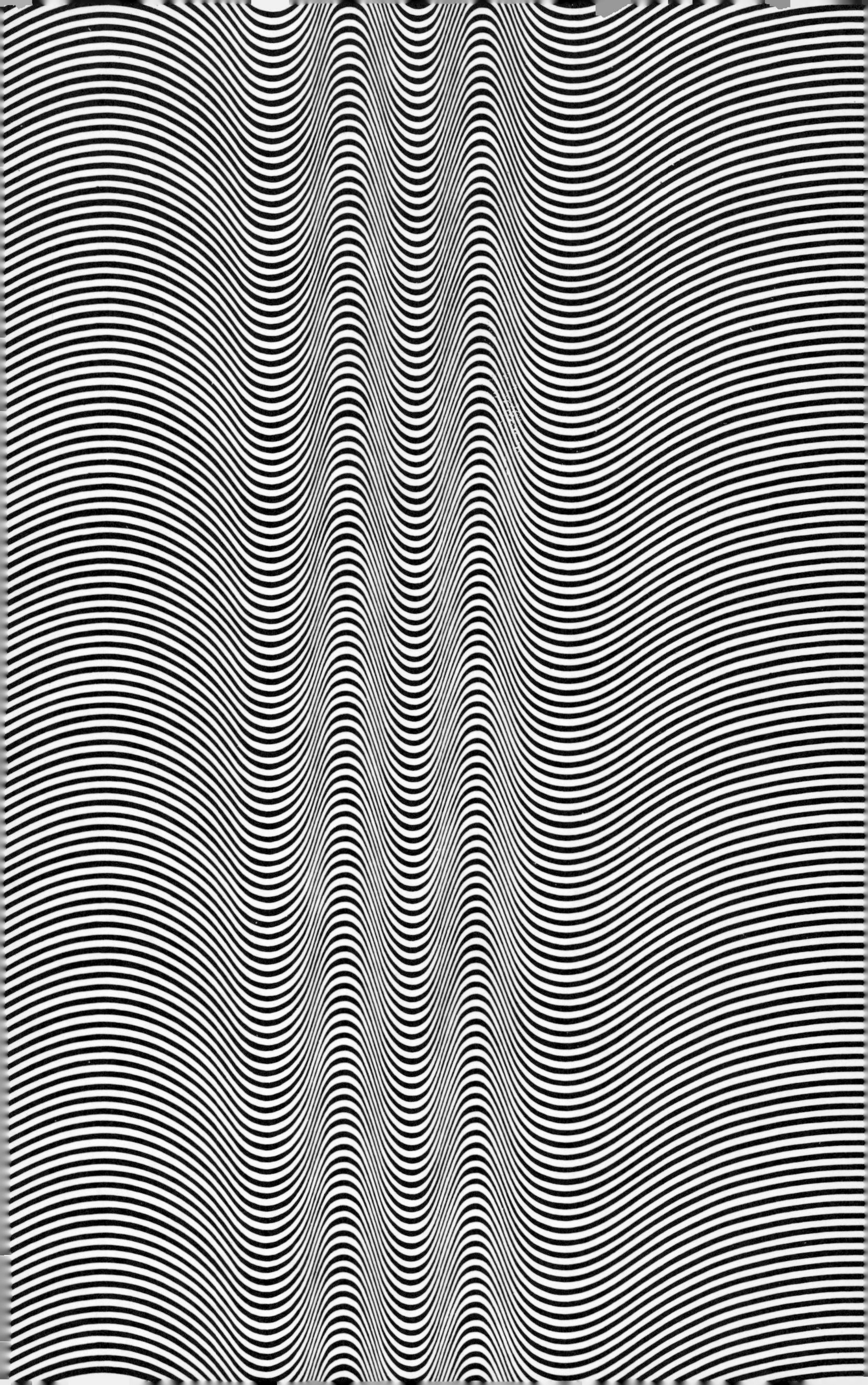

The very essence of nature is rhythm. Rhythm can be found in the regular ebb and flow of the ocean tides, as well as in the alternating cycles of darkness and light that occur every twenty-four hours.

The French astronomer Jean Jacques d'Ortous de Mairan first observed a rhythm in his house plants in 1729. He noticed that the plants closed their leaves in the evening and opened them again each morning. As an experiment, he placed one plant in total darkness for a period of time. He was amazed to discover that the plant continued to close and open its leaves regularly even though it was not exposed to sunlight.

Henri-Louis Duhamel, an engineer and agriculturalist, possessed a scientific mind but did not believe the experiments of de Mairan. How is it possible, Duhamel questioned, for a plant to maintain a regular sleep pattern if the plant does not know whether it is night or day? He decided to duplicate the total darkness experiments of de Mairan. And to make sure that the plant received absolutely no light, he placed the plant in a leather trunk, which he covered with blankets and hid in a closet. Still the plant's leaves closed in the evening and opened in the morning.

Duhamel, more puzzled than ever, now wondered what other stimulus could be responsible for the plant's sleep pattern. The night is much cooler than the day, he reasoned. Could the stimulus be a change in temperature? Duhamel placed his plant in a hothouse and raised the interior temperature well above normal. He recorded the results of this experiment in 1758: "I have seen this plant close up every evening in the hothouse even though the heat of the stoves had been much increased. One can conclude from these experiments that the movements of the sensitive plant are dependent neither on the light nor on the heat."

Scientists studied this strange rhythm over the years. In 1939 Erwin Bünning, a German botanist, published a hypothesis that at first generated little interest. Bünning had determined through studies and experiments with leaf movement that plants have internal circadian (24-hour) rhythms, and these rhythms are used to measure time. In effect, Bünning theorized that plants possess biological clocks. Other scientists of the time thought the Bünning hypothesis too absurd to be believed. Not until many years later, through the accumulated work of other researchers, did evidence point to the possibility of biological clocks in organisms from the lowly one-celled plant all the way to human beings.

The internal world of a human being is influenced by the beat of nature. *Chronobiologists* (biologists who study how the body measures time) have identified four body rhythms in human beings: ultradian rhythm, circadian rhythm, circamensual rhythm, and circannual rhythm.

Ultradian rhythm is a 90- to 100-minute cycle. Studies have shown that daydreaming, levels of concentration, and feelings of hunger rise and fall in ultradian rhythm throughout the day. The sleep pattern is also ultradian in nature.

There are stages of sleep through which one passes during a normal night. As one's eyes close on the world, *alpha waves* are produced in the brain. Alpha waves represent a state of relaxation. The *myoclonic jerk,* a sudden body spasm, may jolt the sleeper awake for a moment. This normal occurrence is the result of a spurt of brain activity.

Slowly one drifts into Stage One, a time of light sleep. Muscles begin to relax, pulse and breathing begin to slow, and body temperature falls. In Stage Two, one's eyes slowly roll from side to side and thoughts are fragmented; one has been asleep about ten minutes. As one descends to Stage Three, muscles are relaxed and body temperature and blood pressure continue to fall. The heart rate slows and breathing is even. One gradually sinks to Stage Four, which is known as *delta sleep.* This is the deepest level of sleep. (The term *delta sleep* is derived from the fact that, at this point, the brain produces large, slow brain waves—delta waves.)

Having reached the bottom of the sleep cycle, one begins to ascend to the stage of light sleep. Instead of one's waking up, though, the eyes begin to dart back and forth.

This period of *rapid eye movement (REM)* is characterized by dreaming. After the REM period ends, one again descends through Stage Two, all the way to delta sleep, and back up again for another interval of REM. This cycle recurs about every 90 minutes during the night.

The name *circadian rhythm* comes from the Latin *circa,* meaning "about," and *dies,* meaning "a day." Body temperature fluctuates one or two degrees Fahrenheit (less than one degree Centigrade) during the 24-hour circadian period. This fluctuation determines if one is a "lark" or an "owl." Larks are people whose body temperature is high in the morning. Larks leap out of bed at the sound of the alarm clock, ready to face a new day. Owls, on the other hand, would prefer to roll over for another forty winks. Their body temperature rises more slowly and peaks later in the day.

Other bodily functions, such as breathing, blood pressure, blood sugar level, and pulse rate, also vary in circadian rhythm. People who travel by plane across many time zones often experience a phenomenon called *desynchronization* (jet lag). Travelers may arrive at their destination just in time for breakfast, but their biological clocks may be telling them that it is time for bed. They may also be fatigued and not very alert for several days while their circadian rhythm adjusts itself to the new time period.

Circamensual refers to the 28- to 30-day rhythm associated with female menstruation. There also seems to be a monthly rhythm associated with males. In the seventeenth century a doctor named Sanctorius began weighing men as an experiment. He found that a monthly one- to two-pound change was evident in the weights of his subjects over a period of time. Modern research has hinted at the possibility of 4- to 6-week cycles of hormonal and mood changes in men also corresponding to the menstrual cycle.

The term *menses* means "lunar month." Experiments at the Rock Reproduction Clinic in Boston have strengthened the theory that menstruation is controlled by the phases of the moon. In the study, women with irregular menstrual periods were asked to sleep in the indirect light of a 100-watt lamp (simulating a full moon) from the fourteenth to the seventeenth nights of their cycle following the onset of menstruation. A gradual regulation of the menstrual period began with ovulation occurring at the time of exposure to the light. Could this regulation of the circamensual rhythm be the birth control method of the future?

The fourth rhythm is *circannual,* or yearly. This rhythm in human beings follows the tempo of the seasonal changes in nature. Children grow rapidly in the spring. The sweltering heat of summer is eased by the secretion of a thyroid hormone. There is evidence that late autumn and early winter are seasons when a man's beard grows rapidly. Robert Sothern, a scientist associated with the Chronobiology Laboratory of the University of Minnesota, thinks that this is related to fur-bearing animals growing thicker coats as they prepare for winter.

Dr. Alain Reinberg of the Rothschild Foundation Hospital in Paris has done studies of human sexual rhythm. He discovered that male hormone levels and sexual activity peak during late autumn and early winter. These results parallel studies of the start of menstruation in girls, which usually occurs at the same time of year. These results might suggest a natural period for human conception.

There also appears to be circannual rhythm for death. Statistics gathered by Professors Michael Smolensky and Franz Halberg of the University of Minnesota and Frederick Sargent of the University of Texas at Houston indicate that deaths from respiratory and heart disease are at their highest point from December through February in both the northern and southern hemispheres. The scientists have also determined that deaths from pneumonia and flu reach their height during a certain period across the United States. This includes those states such as Florida that have mild weather year round and northern states such as Minnesota that have extreme winter and summer conditions.

In explaining the scientists' conclusions, Dr. Alain Reinberg writes: "If the risk of mortality for certain lung diseases is higher between late December and late February—in the northern hemisphere, it is not necessarily because cold and storm weather occurs at this time, but rather because the human organism is then more susceptible to this type of infection than at any other time."

Where inside us is the timepiece that regulates these body rhythms? Scientists don't know for sure. Some chronobiologists are of the opinion that each cell of an organism contains its own internal clock. Other scientists refute the internal clock theory and hold that body rhythms are controlled by external forces such as light and temperature. Still others believe in two clocks—internal *and* external. The theory is that every organism keeps its internal clock from running too fast or too slow by synchronizing with an external force. An exposure to light, such as the 24-hour sun clock, will allow the internal clock to reset itself to be on time with the rest of nature.

This may also be true for the three life cycles—physical, emotional, and intellectual—of the biorhythm theory. The rhythms that begin at birth may be regulated over a lifetime by the 24-hour cycle of light and darkness.

The theory of biorhythm now is controversial. The theory is difficult to prove, and scientists tend to dismiss biorhythm as nonsense.

The fact remains, though, that rhythms exist in nature and in human beings. Years passed before this concept was raised from the level of nonsense to the level of scientific fact. This acceptance was achieved through research.

And it will be research into the theory of biorhythm that will propel this fascinating subject from the cycle of darkness to the cycle of light.

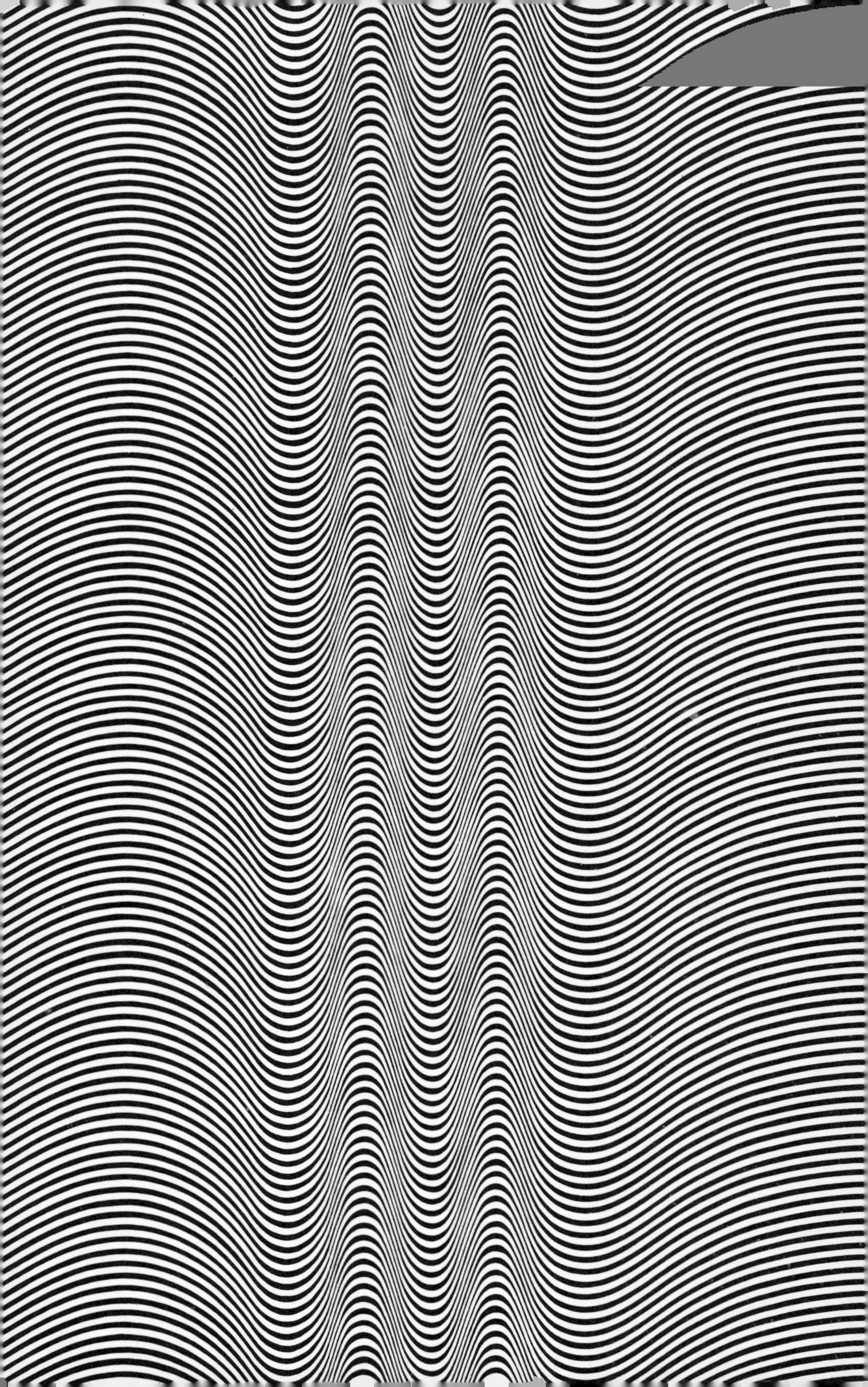

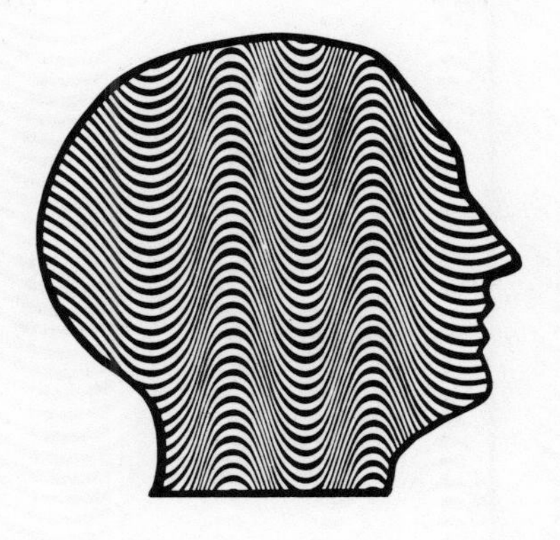

CALCULATING METHODS PAST AND PRESENT

3

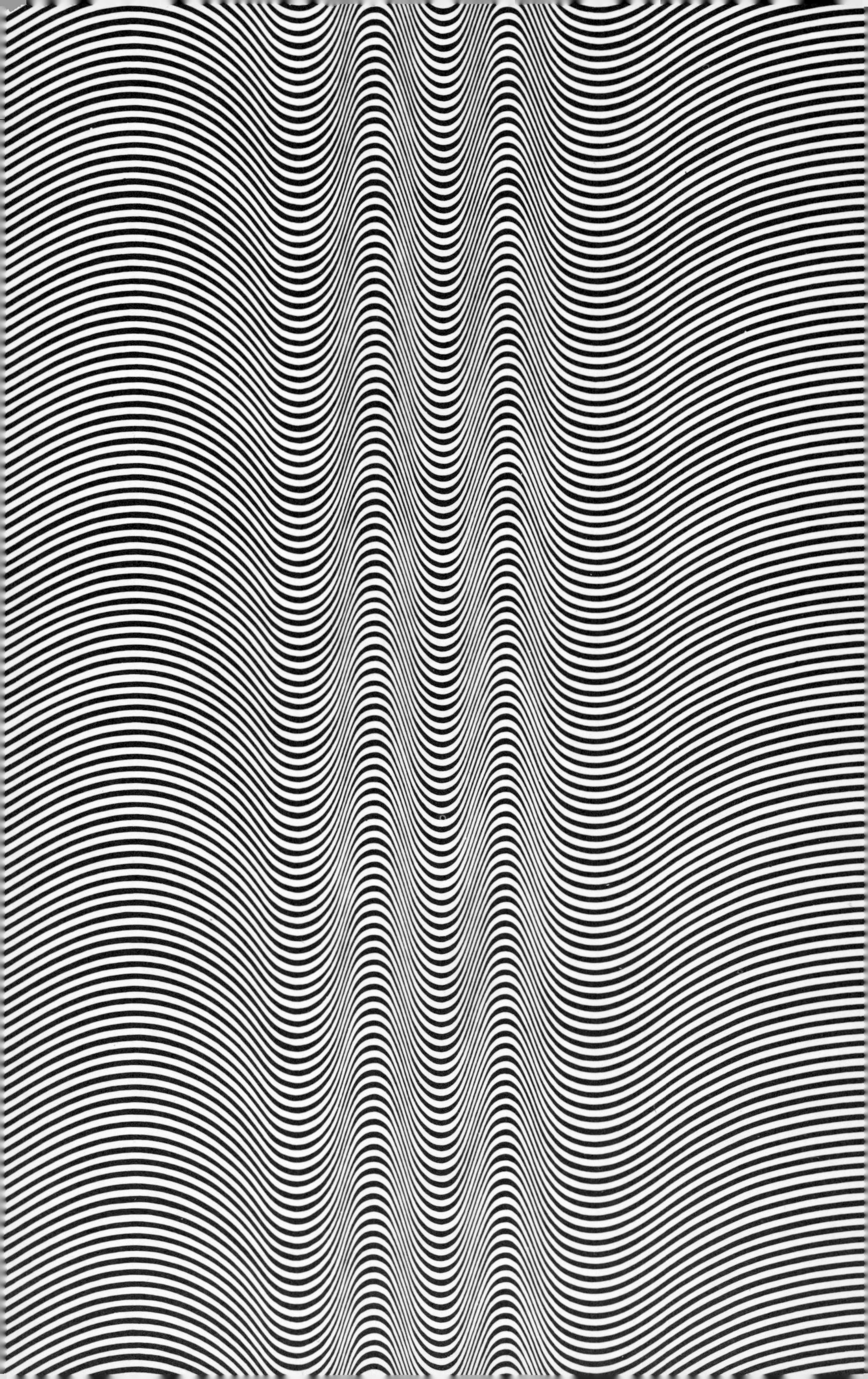

Biorhythm calculating methods began with the lengthy charts and tables of Dr. Wilhelm Fliess of Berlin, and progressed all the way to the computerized biorhythm printout of today. In between is a fascinating chronicle of how biorhythm methods got to here from there.

In 1906 Dr. Fliess published a book about his research into the 23-day and 28-day rhythms. He had hoped his discoveries would turn the world upside down. He was mistaken. The book contained so many complicated calculations, extensive tables, and intricate charts that readers were overwhelmed. There were not many willing or able to decipher the ponderous Fliess formulas. A device was needed to simplify biorhythm computations.

Such a device—a slide rule—was designed between 1904 and 1909 by Dr. Hermann Swoboda of Vienna. The slide rule enabled the user to determine the critical days in the physical and emotional rhythms by use of the birth date. Dr. Swoboda also published an instruction booklet explaining the slide rule's method of calculation. Unfortunately the slide rule was for research purposes and not available to the public. People interested in biorhythm still had to use the lengthy formulas.

Alfred Judt, a German mathematician and doctor of engineering, saw the problems of the complicated biorhythm formulas. He was very interested in evaluating biorhythmically the performances of various athletes. He devised calculating tables based on the athlete's day and year of birth and the date of the athletic event. Although Judt's tables were a step forward, the tables were too specialized to be generally useful.

The calculating tables designed by Judt were revised by Hans R. Frueh, a Swiss engineer and mathematician. He was interested in improving the tables so they could be used more extensively. Besides perfecting the tables, he also invented the hand-operated Biocalculator. One side of the device has the setting for the calculated month; the reverse side displays the calculating tables. This machine, first produced in Switzerland in 1932, is still manufactured today. Frueh also prepared the Bio-Card, a vertical biorhythm chart for individual calculation of the rhythm positions. The Bio-Card uses the colors red, blue, and green to illustrate the plus portions of the physical, emotional, and intellectual rhythms respectively. The minus portions are blank. In that way the critical days can be easily seen.

The general principle of the Bio-Card was adapted in the 1950's to the sine-curve chart. On the sine-curve chart, the plus days are shown above the horizontal zero line, and the minus days are shown below. The rhythms can be illustrated in color—red, blue, and green—or in the international scientific code of solid line, broken line, and dot-and-dash line. Critical days are noted as days when the rhythm crosses the horizontal zero line.

The sine-curve chart is used in the Cyclgraf kit. The kit contains three plastic rulers, calibrated with the days of each rhythm, and calculating tables. The rhythm positions for the first day of the month being calculated are determined from the tables. Each ruler is then positioned on the chart so that the rhythm position for the first day of the month corresponds to day 1 on the chart. The curve of the cycle is then drawn.

Used in much the same way as the Cyclgraf rulers is the three-sided plastic scale developed by the Japanese. Called the Tatai Biorhythm Scale, the curved sides represent the three rhythms. The scale is used on the biorhythm chart to plot the rhythm positions.

The positions of the three rhythms are pictured by the dials of the Dialgraf, a round metal calculator. The first of the six dials is on the outside and shows the days of the week. The second dial has 31 days of the month, with 1 extra day. The next three dials show the three rhythms. The plus days of each rhythm are marked with white numbers on a colored background. The minus days are marked with colored numbers on a white background. The sixth dial controls the month selection. Critical days are easy to see because of the change from color to white. Small triangles also indicate the critical days.

Similar to the Dialgraf is the Japanese Biolucky Disk. The Biolucky is made of four plastic disks. The outside disk represents the numbers of the days of the month plus 2 (from 1 to 33). The three inner disks correspond to the three rhythms and have solid circles indicating plus days and outline circles indicating minus days. Critical days are shown as spade (physical), heart (emotional), and club (intellectual).

Another Japanese development is the Casio Biolator. This pocket-sized calculator accurately computes the three rhythm positions for any day. The year, month, and date to be calculated are entered on the Biolator. The individual's birth year, month, and day of the month are subtracted. When the "bio" key is pushed, three numbers will appear corresponding to the positions of the three rhythms for that particular date. The three numbers can be compared to the biorhythm chart illustrated on the Biolator.

Biorhythm aids have advanced from calculators to watches. The Biostar Electronic Calculator is a Swiss-made watch by Certina. A small window in the face shows three concentric colored bands. Each of the three bands is marked with the number of days in one cycle and passes the window at a different speed. The plus days of the rhythms have colored backgrounds. The minus days have silver backgrounds. The critical days are obvious by the change from colored background to silver background. The Biostar is set to the individual's birth date by a jeweler. Since the device is electronic, battery replacement is required every six months.

The most sophisticated biorhythm charts today are prepared by computer. The Biocom 200 is a desk-top computer recently built by the Japanese. The individual's birth date is entered, along with the month and year to be calculated. Instantly the Biocom shoots out a miniature biorhythm chart for the calculated month. The computer is used extensively throughout Japan by such organizations as traffic police and transportation and insurance companies. Olivetti, the Italian office machine manufacturer, has produced a similar desk-top model called the Microcomputer.

The Bio-Rhythm Report, available through Time Pattern Research Institute in New York City, is printed by the IBM-360 computer. The report has 36 pages of charts projected for a full year, with explanatory text. Edmund Scientific Co. in New Jersey will provide a 12,000-word, 33-page biorhythm report printed by the IBM 370-145. Biorhythm Computers in New York City will also prepare computer charts.

The progression of calculating methods shows that good things can always be made better. Thanks to the ingenuity of human beings, the accuracy of biorhythm is available to anyone at the turn of a dial or the push of a button.

PLUS OR MINUS—WHICH IS IT FOR YOU TODAY?

4

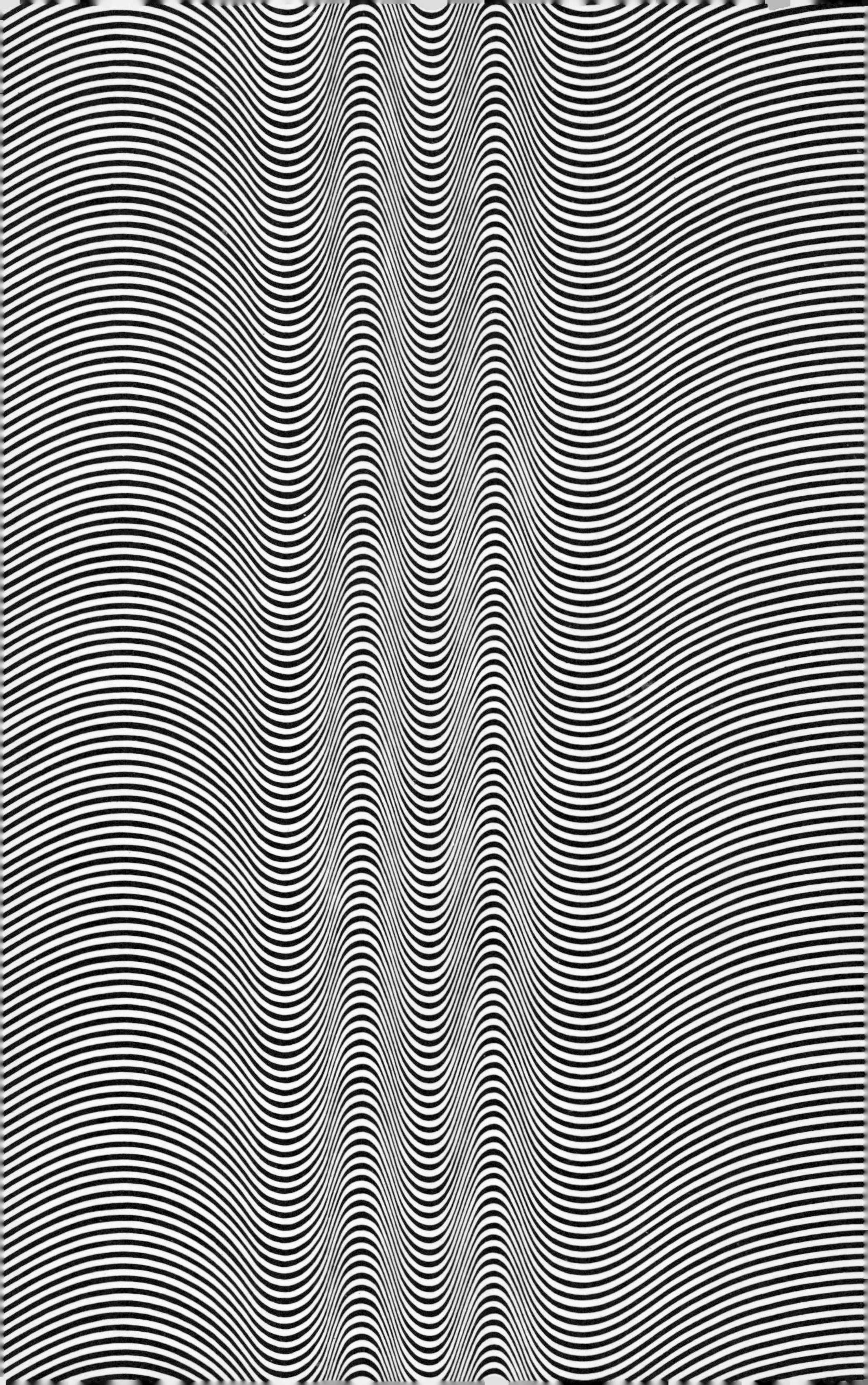

The physical, emotional, and intellectual rhythms begin at the moment of birth and are calculated from that day. The method for determining the positions of the three cycles requires simple mathematics, lined paper, and a calendar.

The total number of days in a person's life are added up from the day of birth to the first day of the charted month. The total is divided by 23, 28, and 33 in three separate divisions. Each of the three divisions will show the number of completed cycles. The remainder will indicate the position of the respective rhythm as of the first day of the charted month.

As an illustration, the following is the calculation of the biorhythm cycles for a fictional person born on May 7, 1963. The charted month will be August 1979.

---

16 years of 365 days (from May 7, 1963, to May 6, 1979)	5,840 days
Extra days for leap years 1964, 1968, 1972, 1976	4 days
Days from May 7, 1979, up to and including August 1, 1979	87 days

May 7 to May 31	25 days
June 1 to June 30	30 days
July 1 to July 31	31 days
August 1	1 day
	87 days

TOTAL	5,931 days

The total number of days in this person's life up to and including August 1, 1979, is 5,931 days. This total is now divided by 23, 28, and 33.

5931 ÷ 23 is equal to 257 completed physical cycles with a remainder of 20 days

5931 ÷ 28 is equal to 211 completed emotional
cycles with a remainder of 23 days

5931 ÷ 33 is equal to 179 completed intellectual
cycles with a remainder of 24 days

The completed calculations show that on August 1, 1979, the individual will be at day 20 of the physical rhythm, day 23 of the emotional rhythm, and day 24 of the intellectual rhythm. As the month begins, therefore, all three rhythms are in the minus phase. New cycles are due to begin on the critical days August 5, August 7, and August 11 respectively. This individual should be more cautious on those days. Notice how each rhythm remains regular through the entire month. The regularity is maintained even as the month changes to September.

"Okay, prove to me that biorhythm works." This challenge is often heard from skeptics who refuse to test the accuracy of biorhythm. Their challenge, however, is misdirected, for the burden of proof lies with each individual. Industrial data, historical documentation, and accident statistics can be quoted to the skeptic ad infinitum, but it is only through personal experience with charting biorhythms that one can be convinced of the theory's validity. You must prove to yourself that biorhythm works, and this is possible in a number of ways.

One way is to draw up biorhythm charts for two or three months in advance. Put the charts away in a desk drawer for safekeeping. Then make notations of how you feel every day. Are you short-tempered or easygoing one day? Are you energetic or lethargic the next? After three months, pull out the biorhythm charts to see how the cycles compare with your notes on self-observation.

Your daily record will be especially helpful when you examine your emotional rhythm. The emotional rhythm is the only cycle in which the critical days (first and fifteenth) fall on the same weekday—the day of birth. A person born

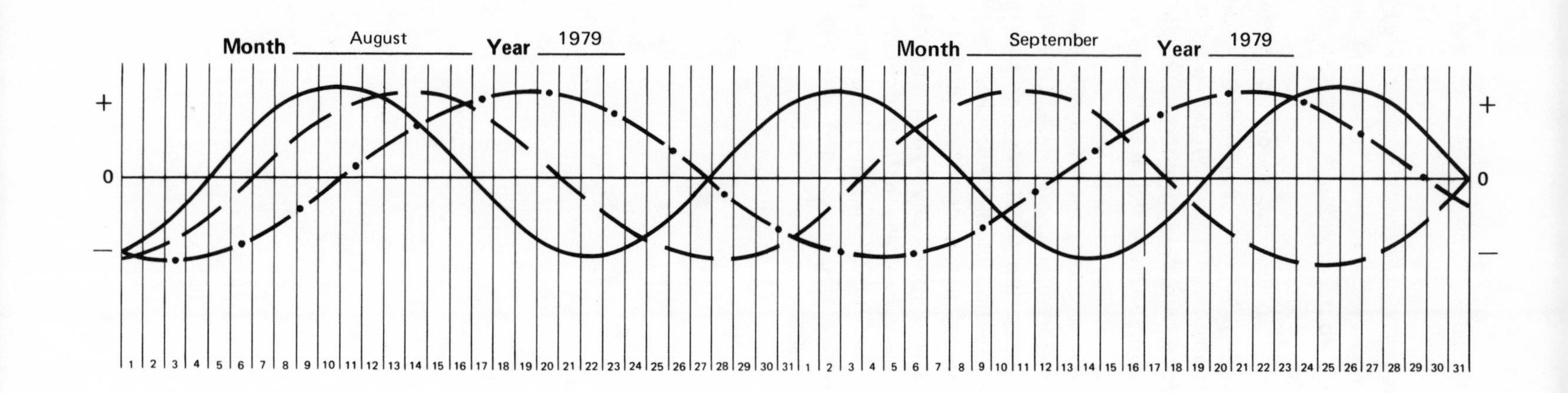
Month August Year 1979
Month September Year 1979
+
0
−
1 2 3 4 5 6 7 8 9 10 11 12 13 14 15 16 17 18 19 20 21 22 23 24 25 26 27 28 29 30 31
1 2 3 4 5 6 7 8 9 10 11 12 13 14 15 16 17 18 19 20 21 22 23 24 25 26 27 28 29 30 31

on a Wednesday can expect a critical day every other Wednesday. You can, therefore, check to see if a pattern exists in which your emotionally unstable days fall on the same day of the week.

Another way to test the theory is to select a date in your past on which an error or accident occurred for which you were responsible. By comparing the date of that personal experience with the biorhythm chart for that day, you may well find that you were not at your best at that particular time.

The rewards are many for those who examine biorhythm in the light of personal experience. Open-minded individuals will recognize the soundness of the theory and will gain an understanding of the inner self by utilizing the information contained in the monthly charts.

For those who reject biorhythm without first exploring the concept for themselves, there is only the advice of that farsighted patriot Patrick Henry: "I have but one lamp by which my feet are guided, and that is the lamp of experience."

INDUSTRY, ACCIDENTS, AND BIORHYTHM

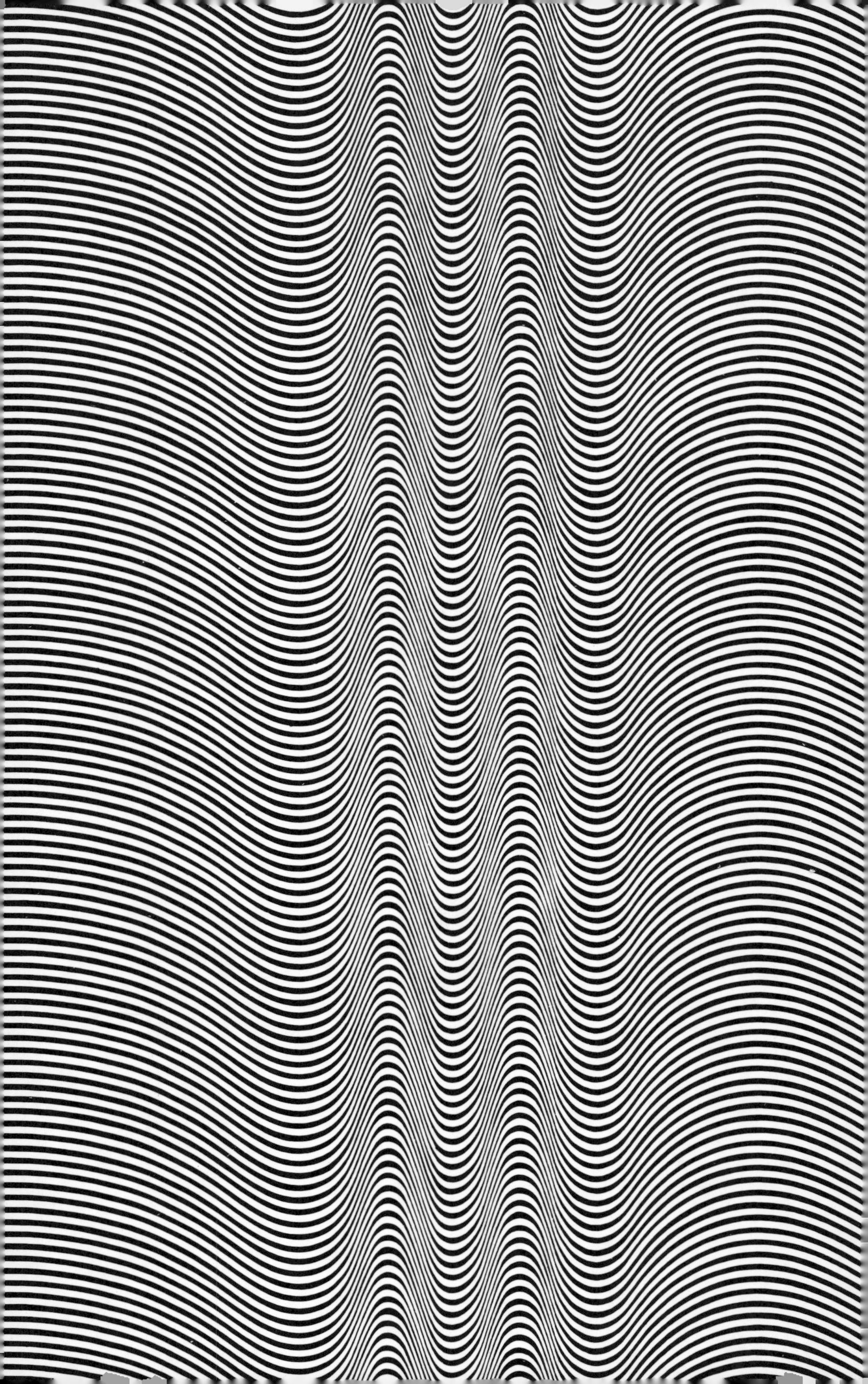

Biorhythm has been used effectively in the United States and abroad in reducing accident rates. The most interesting results have been achieved in Japan and Switzerland.

Switzerland has used biorhythm for many years. In Zurich the Municipal Transit Company warns its trolley and bus drivers of approaching critical days. By making the employees more aware of days of instability and greater potential for accidents, the accident rate per 10,000 kilometers has been reduced by 50 percent.

The airline company Swissair has been examining the biorhythm cycles of its flight personnel for more than ten years. Pilots and co-pilots are not allowed to fly together if both are experiencing critical days at the same time. Accident rates have been sharply reduced.

In Japan, biorhythm was introduced through the efforts of the Japan Biorhythm Association. The response to the theory was overwhelming. Within a few years over five thousand companies and businesses had adopted the use of biorhythm for their employees. The Seibu Transport Company was concerned about the mistakes made in the computerized filing system by its key punch operators. Biorhythm charts were drawn. On a critical day the operator would be switched from key punching to another department. Mistakes were reduced by 35 percent within six months. The Meiji Bread Company began to warn its drivers to exercise caution on critical days. The first year saw a 45 percent reduction in vehicular accidents.

The Omi Railroad, despite its name, is a bus and taxi company operating in the heavily populated cities of Kyoto and Osaka. An initial examination of the company's safety record showed that, of 331 accidents within a five-year period, 59 percent had occurred during a critical period in the driver's rhythms. In 1969 the company began issuing warning cards to drivers on critical days. The accident rate plummeted by 50 percent during the first year. By 1973 the use of biorhythm had allowed the bus division to achieve an accident-free record in over 4 million traveled kilometers.

The Tokyo Metropolitan Police conducted a study of biorhythm in relation to traffic accidents. The study, which was published in 1971, indicated that 82 percent of reported accidents in 1970 had occurred on the driver's critical day. The results of this study led the Traffic Department of the police, together with the Japan Biorhythm Association, to initiate defensive driving programs throughout Japan.

Classes are held to teach drivers about biorhythm and to explain the method for drawing biorhythm charts. Driver safety courses are sponsored by insurance companies. In twelve districts, drivers regularly receive biorhythm charts. Other sections provide charts to those getting or renewing a driver's license and to all persons involved in accidents. The various programs have been highly successful in making the Japanese accident rate drop. The accident rate fell because people became more aware of their biorhythm condition on a given day. Clever methods have been introduced to foster this awareness.

Some companies include biorhythm charts with the employee's weekly pay envelope. On critical days the employee may receive a warning card to be extra careful. Bus and taxi companies distribute origami cranes (the Japanese symbol of happiness and good fortune) to their drivers. The cranes are made out of folded colored paper, the color corresponding to the driver's biorhythm condition. With the crane perched on the dashboard, the driver will not forget to exercise caution during the day. Motorcyclists who deliver mail and telegrams use a similar color system. Colored triangular flags are attached to their motorcycles during critical days. The flag is a critical-day reminder to the driver and also serves to warn motorists and pedestrians.

Another unique system utilizes color-coded empty candy boxes. Some taxi companies provide the boxes to their drivers at the beginning of each critical day. If the driver has had no accidents during his shift, he can turn in the empty box at the end of the day and receive a full box of candy.

While the acceptance of biorhythm has been dramatic in Switzerland and Japan, this is not the case in the United States. Very few companies are willing to declare their interest in biorhythm openly. Many businesses are taking a "wait and see" attitude toward the theory. But there are organizations inclined to view biorhythm favorably. One of these is NL Industries (formerly known as National Lead Company).

An experiment was conducted in the titanium division of the company in July 1965 to determine if biorhythm could reduce the spiraling accident rate. The experiment involved the workers in the rigging department, the millwright shop, and the pipe shop. Biorhythm charts were drawn up for each employee in the rigging and millwright departments. The biorhythm information was given to the respective foremen. When any employee in the two departments had a critical day, the foreman would stress the importance of safety, give more supervision, and substitute less hazardous jobs.

The pipe shop was used as the control group. No biorhythm charts were done for those workers, although they did receive the same safety information as the other two departments.

While the results of the experiment don't prove conclusively that biorhythm works, company officials were impressed. During the last five months of 1965 the rigging department reduced its accident rate 18 percent. In 1966 the rate was cut 42 percent for the riggers, while the millwright shop achieved a 4 percent reduction.

How did the pipe shop do?

The pipe shop, which received no biorhythm information, had an injury rate *increase* of 28 percent.

Skeptics might say that *any* program in which workers were given individual attention would improve morale and so reduce the accident rate. But the Executive Safety Committee was convinced that the biorhythm program was responsible.

United Airlines noticed a large decrease in the number

of accidents through biorhythm experiments. Computerized biorhythm charts were done for interested ground crew and maintenance employees. The biorhythm information was provided to foremen, who advised workers to be extra careful on critical days. Although safety was discussed, no job reassignment was done. Between 1973 and 1974 accidents were reduced by more than 50 percent. United maintains, however, that the program is on a voluntary basis, strictly experimental and not a company policy.

Bell Telephone Laboratories is interested in the biorhythm theory. The company's hope is that the use of biorhythm will increase productivity of all employees, including engineers and scientists. The Human Technology Division will be conducting the research. The actual work is still in the beginning stage.

There are many companies that deny any experimentation with biorhythm. Perhaps they feel the theory is too controversial. Possibly the companies want more documentation that biorhythm works. If the latter is the case, it is gratifying to know that two research laboratories in the United States are conducting research on biorhythm theory.

One research center is the Man-Machine Design Laboratory of the Naval Postgraduate School in Monterey, California. The experiments of Dr. Douglas E. Neil have been concerned with examining the relationship of accidents to the total biorhythm cycle—plus days, minus days, mixed days—not just critical days. His accident analysis has suggested that accidents do not occur at random. There are certain periods when accidents occur more often. He also stresses that the statistical rate for accidents on a critical day is very high.

The other research laboratory is headed by Harold R. Willis of the Department of Psychology at Missouri Southern State College. Mr. Willis has done some intriguing work concerning fatal one-car accidents. Data were studied on one-car accidents in Missouri that were fatal to the driver. Of the 100 cases presented, the following percentages were found:

46%	occurred on a critical day
11%	occurred within 24 hours of a critical day
57%	total

The results are pertinent because 57 percent of the deaths involved a driver with an emotional critical day or a double critical day (emotional critical plus either physical or intellectual critical). This would seem to indicate that the emotional cycle—which influences judgment—had a great effect on accidents.

Accident research must continue if biorhythm is to be as widely accepted in the United States as it has been in such countries as Switzerland and Japan. Industry must be convinced that biorhythm can reduce accident and injury rates. By accepting biorhythm, both industry and labor stand to gain. Industry can cut losses, and labor will be assured of a safer working day. Safety for the American worker should be the prime concern of industry. Biorhythm may well be a step in the right direction.

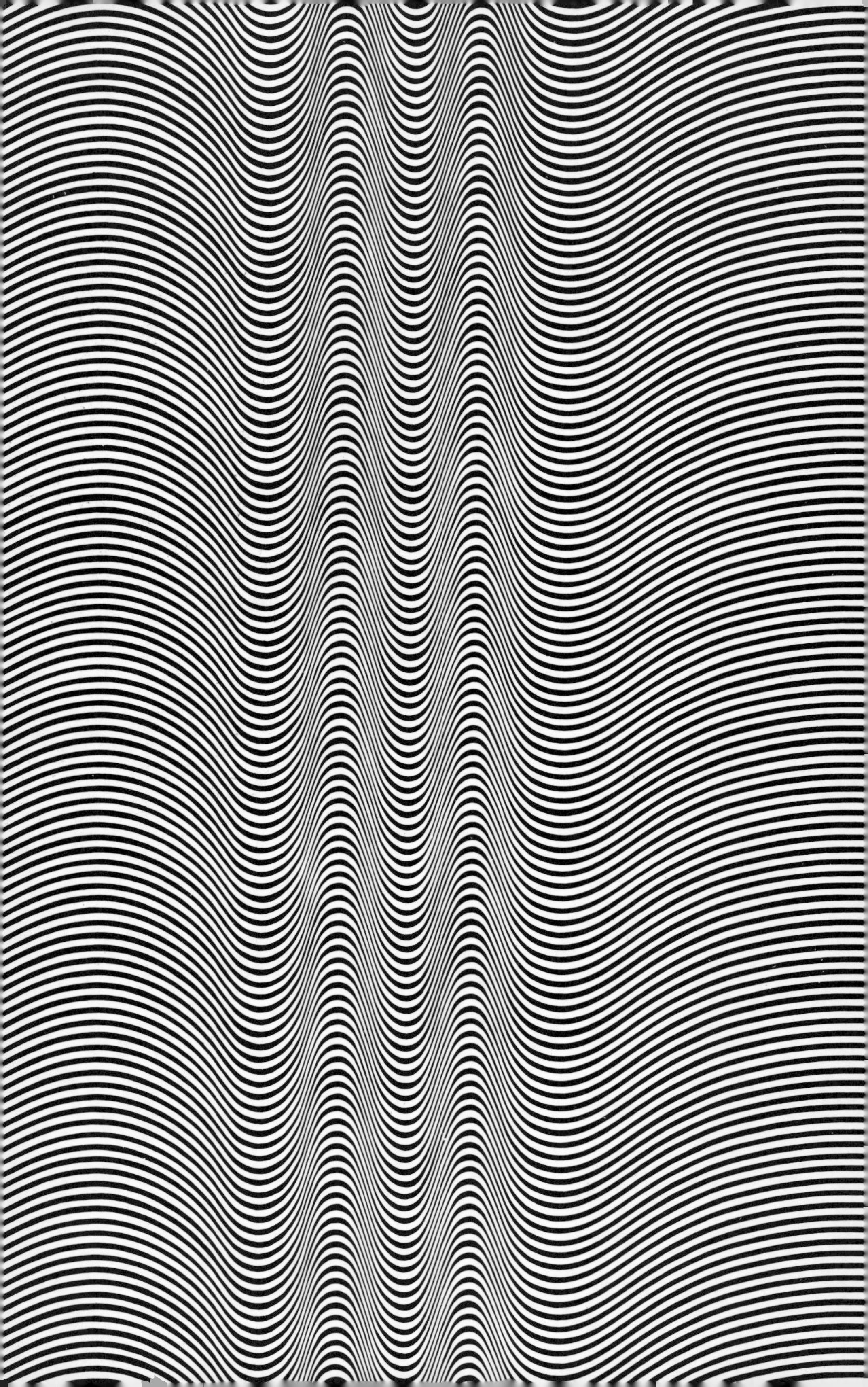

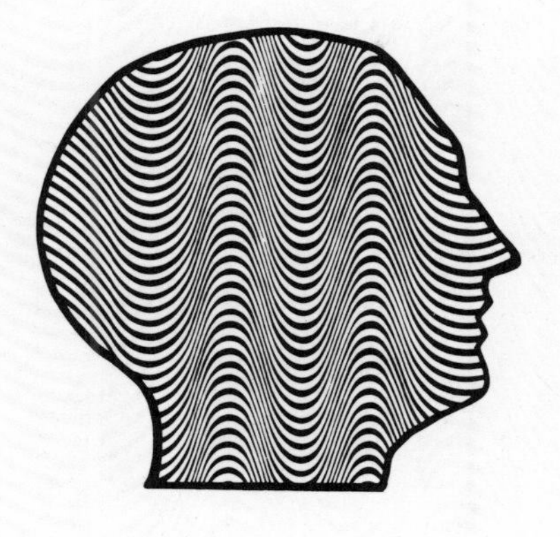

GAZE INTO THE CRYSTAL BALL

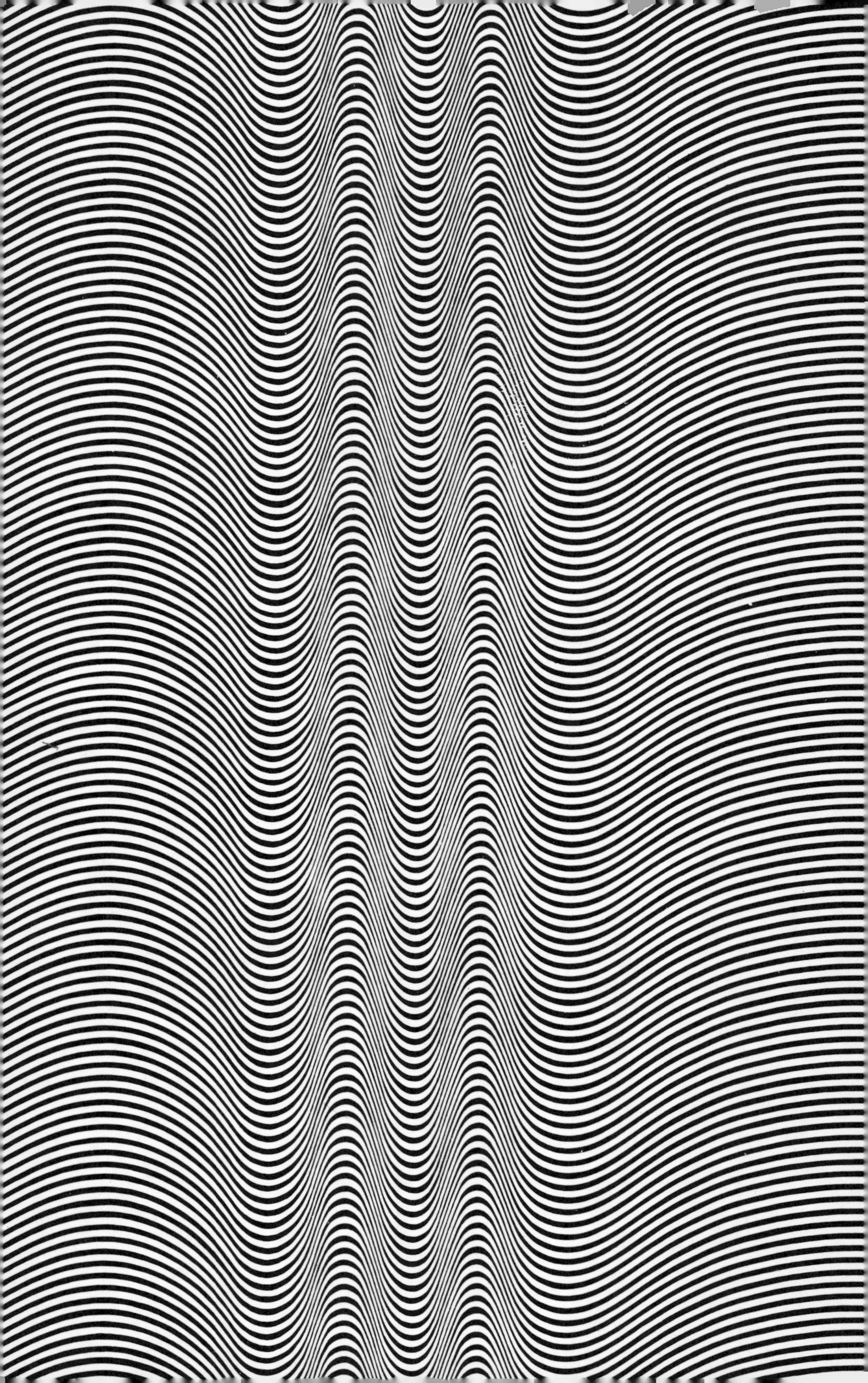

For I dipt into the future, far as human eye could see,
Saw the Vision of the world and all the wonder that would be.

—Alfred, Lord Tennyson: "Locksley Hall"

The biorhythm world of the future—what will that world be like?

Biorhythm researchers are giving hints about the future through the investigation of such concepts as measuring the compatibility of two persons, advance determination of the birth date and sex of an unborn child, and life centered around biorhythmic planning. These concepts hold exciting promise for the years ahead. Let's dip into the world of tomorrow to see the potential for biorhythm in our lives.

The day will come when each person will consult the biorhythm chart each morning just as one checks the daily weather report. The individual will be able to schedule the day's activities according to the biorhythm "forecast."

In schools, students with similar plus and minus phases can be grouped together. New subjects can be introduced during plus intellectual periods when learning potential is high and mental response is keen. Practice and review can be scheduled for minus days.

Industry can reduce accidents and injuries by shuffling employees to less hazardous tasks on critical days. Productivity can be increased if challenging jobs are assigned to workers during plus phases. Better working relations are possible through worker compatibility studies.

In medicine, doctors can avoid performing delicate surgery on their critical days and can also schedule operations for the patient's best physical period. Patients will have a speedy recovery if recuperation occurs during the minus phase of their rhythms. Victims of heart attacks can be given more attention and care on critical days—the days on which most deaths occur.

Athletes can plan training for their physical plus days. Individual sports contests can be scheduled for the best days. Compatibility studies can bring the best players together in team sports.

Great things are possible for the person who follows the natural rise and fall of the body's rhythms. A life of biorhythmic planning will allow the individual to realize full potential and achieve harmony of mind and body.

Harmony is also the goal in our relationships with others. In striving for harmony, our compatibility with those we live and work with is an important consideration. Compatible persons get along better and work well together with little friction. Others, unfortunately, always seem to be out of sync with those around them.

Is there any way to determine a potential problem of divergent dispositions?

One way may be through biorhythm compatibility study. Biorhythm studies have indicated that there seems to be a greater degree of compatibility in people who have a similarity in their biorhythm cycles. This similarity would mean that the plus and minus phases of one person would coincide with the plus and minus phases of another. For example, biorhythm curves would be exactly the same in two people sharing the same birth date. These two people would have 100 percent biorhythmic compatibility in all three rhythms.

Sharing similar rhythm patterns, though, does not guarantee a successful relationship at all times. In the same way, a low degree of compatibility does not doom a relationship from the start. Each of us is a unique individual, vulnerable to the stresses of every new day. Through biorhythm compatibility studies an individual can recognize the changing temperaments and feelings of those around him or her, and use the insight gained to ride out the stormy weather of an incompatible period.

Take the example of a "lark" and an "owl." The lark is up at the sound of the alarm clock, raring to go, while the owl finds it a monumental task just to throw off the bed covers. One doesn't need a crystal ball to see how these two incompatible types would clash in the morning. By recognizing their incompatibility, discord can be avoided. Perhaps the lark could sing in the shower rather than at the breakfast table in deference to the owl. Incompatibility tempered with love and understanding can lead to a harmonious relationship.

How is biorhythm compatibility determined?

Compatibility depends on the number of days that separate the rhythms of two or more persons. Begin by calculating the positions of each person's rhythms for the same month. The following is a comparison of the rhythms of two friends, Jack and Roger, for the month of October 1979.

CHART A

JACK'S BIRTH DATE: JULY 1, 1965

14 years of 365 days (from July 1, 1965, to June 30, 1979)		5,110 days
Extra days for leap years 1968, 1972, 1976		3 days
Days from July 1, 1979, up to and including October 1, 1979		93 days
July 1 to July 31	31 days	
August 1 to August 31	31 days	
September 1 to September 30	30 days	
October 1	1 day	
	93 days	
TOTAL		5,206 days

5206 ÷ 23 is equal to 226 completed physical cycles with a remainder of 8 days

5206 ÷ 28 is equal to 185 completed emotional cycles with a remainder of 26 days

5206 ÷ 33 is equal to 157 completed intellectual cycles with a remainder of 25 days

ROGER'S BIRTH DATE: SEPTEMBER 26, 1965

14 years of 365 days (from Sept. 26, 1965, to Sept. 25, 1979)	5,110 days
Extra days for leap years 1968, 1972, 1976	3 days
Days from September 26, 1979, up to and including October 1, 1979	6 days

September 26 to September 30	5 days
October 1	1 day
	6 days

TOTAL	5,119 days

5119 ÷ 23 is equal to 222 completed physical cycles with a remainder of 13 days

5119 ÷ 28 is equal to 182 completed emotional cycles with a remainder of 23 days

5119 ÷ 33 is equal to 155 completed intellectual cycles with a remainder of 4 days

Jack begins the month of October 1979 with his physical rhythm at day 8, his emotional rhythm at day 26, and his intellectual rhythm at day 25.

Roger begins the same month with his physical rhythm at day 13, his emotional rhythm at day 23, and his intellectual rhythm at day 4.

To determine the number of days separating each rhythm, subtract the smaller number from the larger number.

MONTH: OCTOBER 1979

	Physical	*Emotional*	*Intellectual*
Jack	8	26	25
Roger	13	23	4
	5	3	21

The figures 5, 3, and 21 are the number of days separating the physical, emotional, and intellectual rhythms, respectively, of Jack and Roger. Compare these figures to the biorhythm compatibility chart to determine to what degree the friends are compatible.

CHART B

BIORHYTHM COMPATIBILITY CHART

	Excellent	*Good*	*Average*	*Fair*	*Poor*
Physical	0–2 23–21	3–4 20–19	5–6 18–17	7–9 16–14	10 13 –11½
Emotional	0–2 28–26	3–5 25–23	6–8 22–20	9–11 19–17	12 16 –14
Intellectual	0–3 33–30	4–6 29–27	7–9 26–24	10–13 23–20	14 19 –16½

The chart shows that Jack and Roger have average compatibility in the physical rhythm. They would probably do well together in physical activities such as bicycling; but because their plus and minus phases occur at different times, they might not work well together in team sports.

Good compatibility exists in the emotional rhythm. Jack and Roger would tend to get along well because they would experience the ups and downs of mood changes at the same time. This rhythm is important for members of a family, especially if a degree of incompatibility is present. Family members can work on improving relationships by recognizing the inherent incompatibility in their emotional rhythms.

Jack and Roger have fair compatibility in the intellectual rhythm. This rhythm, which influences learning ability, is not as important as the physical or emotional rhythms in determining compatibility.

Biorhythmic compatibility is in the experimental stage. There is still much to be learned about why some people get along while others do not. Somewhere in the future may lie the answer to this intriguing question of human compatibility.

Another interesting facet to the study of biorhythm in the future is the concept of predicting the day of birth. Birth is a traumatic event. For 280 days, the unborn child exists within the warm protection of the mother's uterus. Suddenly uterine contractions begin. The child is pushed and squeezed through the birth canal. The outside world is cold, noisy, and totally unlike the security of the womb. Is it any wonder that the day of birth is a triple critical day for the child?

Of course, birth is as much a physical shock for the mother as it is for the child. Dr. Hermann Swoboda, one of the pioneers of biorhythm, realized this and wondered how the day of a child's birth could be related to the mother's biorhythm cycles. Through the analysis of many family trees, he was the first to recognize that births followed a regular pattern. He found that births usually occurred on or near the physical or emotional critical day of the mother.

How is the date of birth predicted?

The average human gestation period is 9 months, or 280 days, from conception. Two hundred and eighty days are added to the date of conception to determine the date of expectancy. The biorhythms of the mother are calculated for the birth month, and the critical day closest to the day of expectancy is selected as the projected birth date.

Sometimes the date of conception is not known. In that case, simply calculate the mother's biorhythms for the birth month and select the critical day closest to the date of expectancy.

At best, the projected date of birth is an educated guess. One hundred percent accuracy is not always possible for such a complicated, variable life process such as birth. But the use of biorhythm to determine the date of birth has had some interesting results.

The method can be illustrated with a fictitious mother-to-be whose birth date is September 13, 1954. The date of conception is January 22, 1979. The first step is to project forward 280 days to the date of expectancy.

January	9 days
February	28 days
March	31 days
April	30 days
May	31 days
June	30 days
July	31 days
August	31 days
September	30 days
October	29 days
	280 days

The date of expectancy is October 29, 1979. Next calculate the biorhythm cycles of the mother-to-be for the month of October 1979.

25 years of 365 days (from Sept. 13, 1954, to Sept. 12, 1979)	9,125 days
Extra days for leap years 1956, 1960, 1964, 1968, 1972, 1976	6 days

Days from September 13, 1979, up to and including October 1, 1979		19 days
September 13 to September 30	18 days	
October 1	1 day	
	19 days	
TOTAL		9,150 days

9150 ÷ 23 is equal to 397 completed physical cycles with a remainder of 19 days

9150 ÷ 28 is equal to 326 completed emotional cycles with a remainder of 22 days

---

The physical rhythm would be at day 19 and the emotional rhythm would be at day 22 on October 1, 1979. The critical days in the physical rhythm would be 6, 17, and 29. The critical days in the emotional rhythm would be 8 and 22. In this instance, the closest critical day and the day of expectancy would be the same day. The projected day of birth is, therefore, October 29.

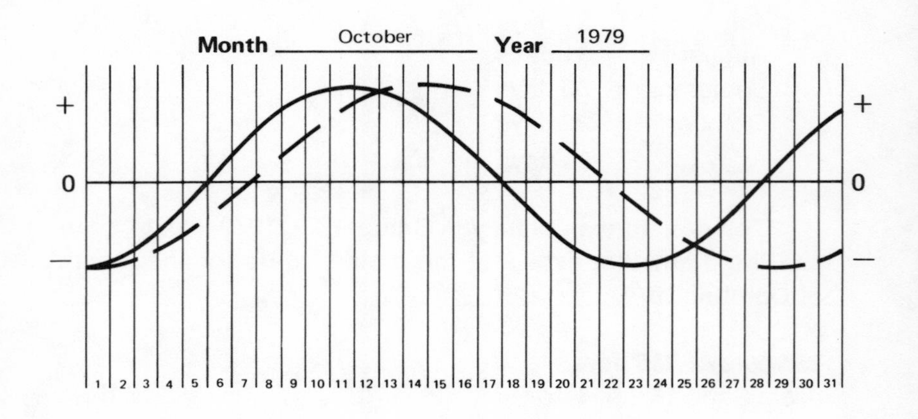

Biorhythm chart for mother-to-be born September 13, 1954.
Projected date of birth is October 29, 1979.

Other examples can be used to illustrate this interesting concept further.

Jacqueline Kennedy, born July 28, 1929, gave birth to a daughter, Caroline, on November 27, 1957, which was one day *before* a critical day in Mrs. Kennedy's physical rhythm.

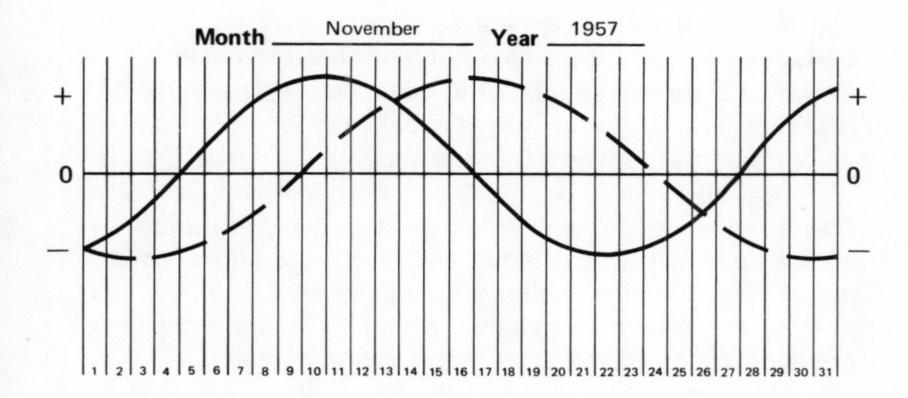

Jacqueline Kennedy gave birth to a son, John, Jr., on November 25, 1960, which was one day *after* a critical day in Mrs. Kennedy's physical rhythm.

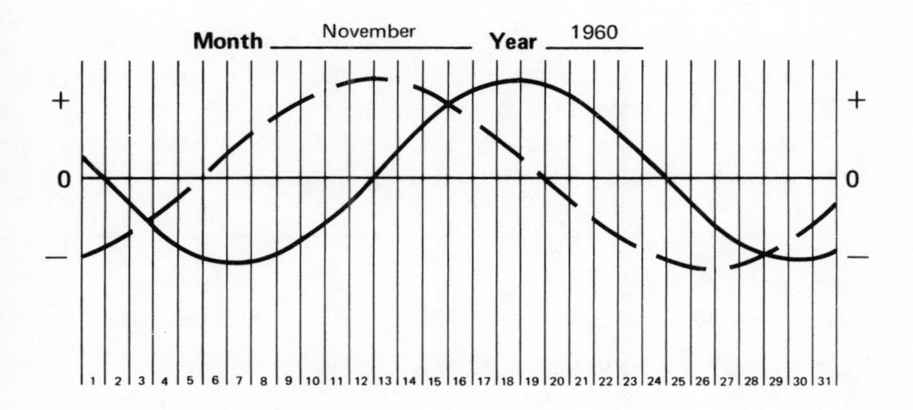

In addition to predicting the date of birth, biorhythm can also be used to predict the sex of the unborn child.

Sex is determined at the time of conception. If the ovum is penetrated by a sperm with an X chromosome, the fertilized egg will be female. If the sperm has a Y chromosome, the egg will be male. The father, then, determines the sex of the fertilized egg. Some biologists, however, believe that the egg is predisposed to accept one sperm over the other at certain times. Biologists also contend that the alkalinity or acidity in a woman's blood is a factor in sex determination. Alkalinity is associated with the conception of males and acidity with the conception of females.

The biorhythmic approach to sex determination holds that a high physical rhythm at the time of conception increases alkalinity in the blood and favors the egg's acceptance of the Y sperm (male). A high emotional rhythm at the time of conception increases the acidity in the blood and predisposes the egg to accept the X sperm (female).

The time of conception is the key to the determination of sex. Conception must occur when one of the two rhythms is clearly high. If both rhythms are in the plus phase simultaneously, a definite prediction cannot be made.

A prediction is made by obtaining the date of expectancy and subtracting 280 days. This will determine the possible date of conception. Calculate the mother's biorhythms for the month and locate the positions of the two rhythms for the time of conception.

John Kennedy, Jr., was born on November 25, 1960. Projecting back 280 days gives the date of February 18, 1960, as the possible date of conception.

Mrs. Jacqueline Kennedy was born July 28, 1929, and her biorhythms must be calculated for February 1960.

---

30 years of 365 days (from July 28, 1929, to July 27, 1959)	10,950 days
Extra days for leap years 1932, 1936, 1940, 1944, 1948, 1952, 1956	7 days

Days from July 28, 1959, to February 1, 1960		189 days
July 28 to July 31	4 days	
August 1 to August 31	31 days	
September 1 to September 30	30 days	
October 1 to October 31	31 days	
November 1 to November 30	30 days	
December 1 to December 31	31 days	
January 1 to January 31	31 days	
February 1	1 day	
	189 days	
TOTAL		11,146 days

11,146 ÷ 23 is equal to 484 completed physical cycles with a remainder of 14 days.

11,146 ÷ 28 is equal to 398 completed emotional cycles with a remainder of 2 days.

The physical rhythm would be at day 14, and the emotional rhythm would be at day 2 on February 1, 1960. Transferring this information onto a biorhythm chart clearly shows the physical rhythm to be high for the date of February 18, favoring the conception of a male child.

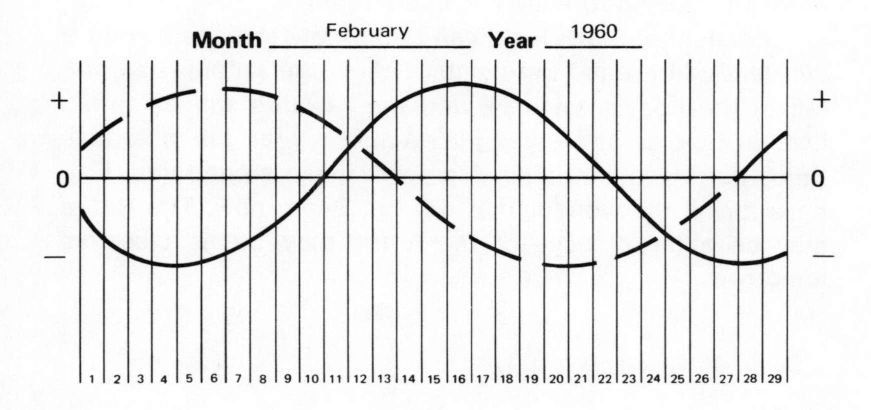

Cher Bono, born May 20, 1946, gave birth to a daughter, Chastity, on March 4, 1969. Projecting back 280 days gives the date of May 28, 1968, as the possible date of conception. Cher's biorhythm chart for May 1968 shows the emotional rhythm to be high for the twenty-eighth, favoring the conception of a female child.

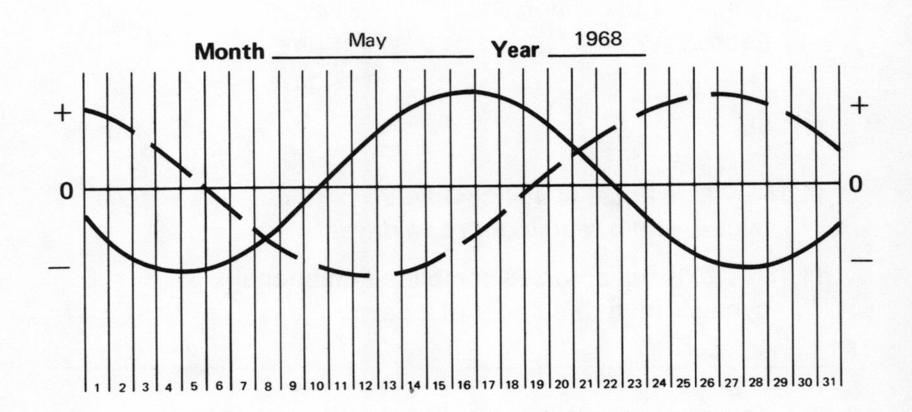

Biorhythmic prediction of the sex of an unborn child is not 100 percent accurate at the present time. As with the other concepts discussed, such accuracy can only come when more is learned about the influence of biorhythm in our lives. This knowledge lies in the future.

Farsighted individuals can begin now to benefit from a life centered around biorhythm. Follow the natural rise and fall of the body's rhythms, and the rewards will be great. Live a happier and safer life. Achieve your full potential. Enjoy harmonious relationships with those around you. Prepare for all the wonder that will be. Begin now. The future may be a far-off day—or the future may be as close as tomorrow.

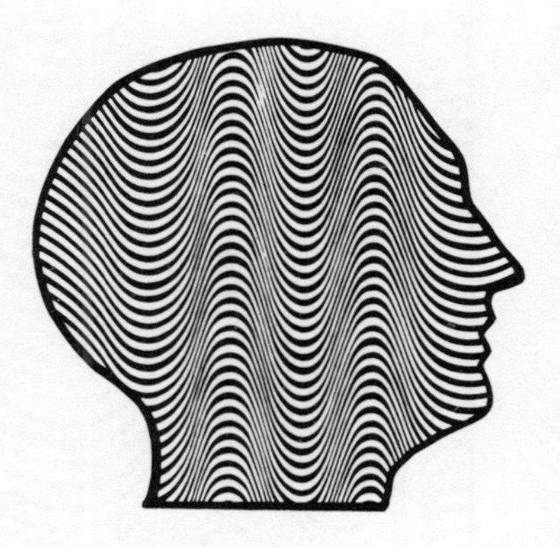

BIORHYTHM AND YOU

7

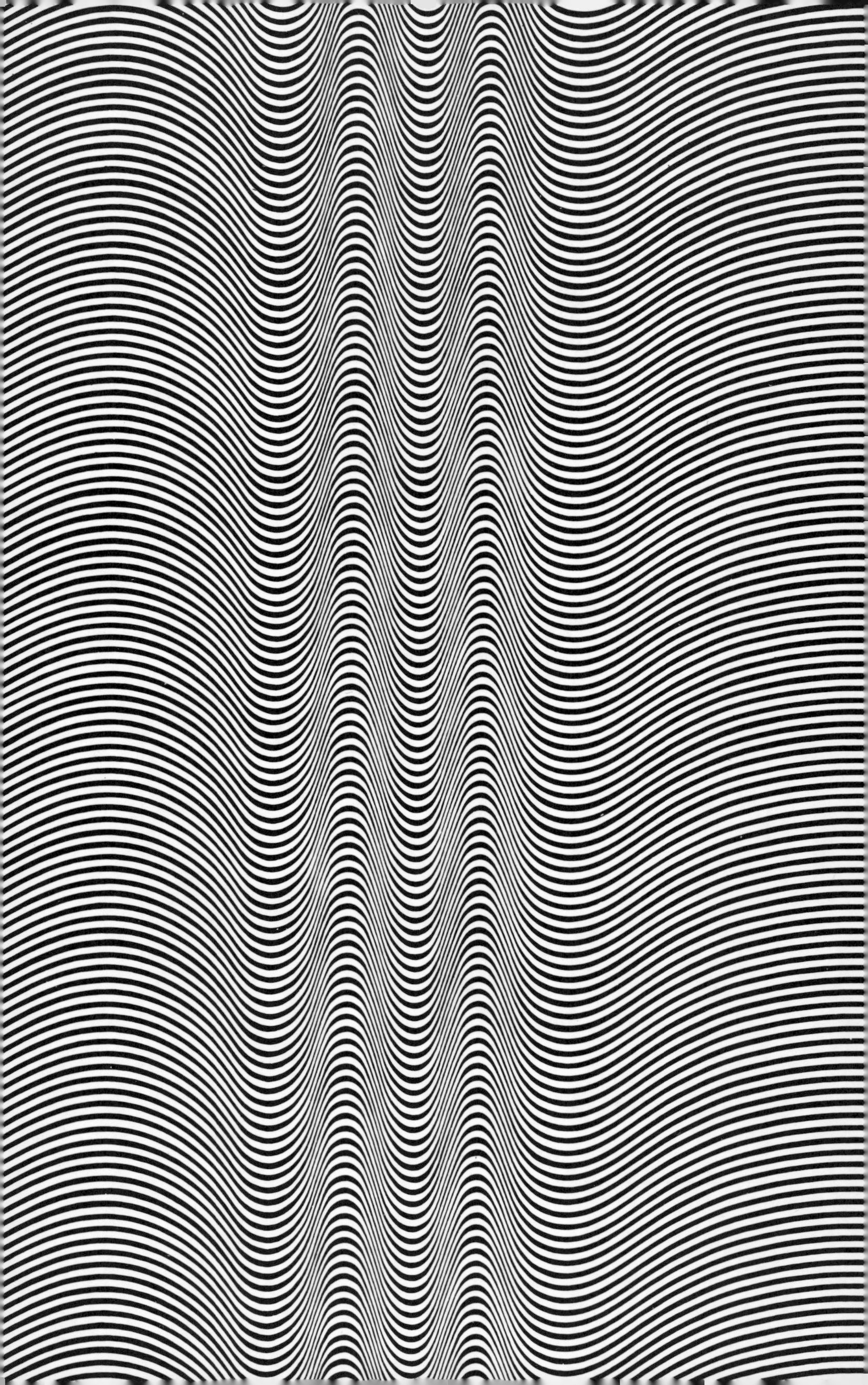

Why not change your life right now? Take a few minutes to calculate your biorhythms. Find out exactly where the three rhythms are positioned for today.

Are you experiencing a physical plus day or an emotional minus day? Maybe today is a critical day in the intellectual rhythm.

Whatever your biorhythmic condition, you know more about yourself now than you did a few minutes ago. The cycles may explain why it was so easy making that basketball foul shot in gym class or why you were so irritable with your best friend.

Now look at the rhythm positions for tomorrow. Use this knowledge about yourself to make tomorrow a better day in school, on the job, on the sports field, and during leisure time. Make the most out of each day and each activity by letting biorhythm touch every area of your life.

Biorhythm can help you do better in school. Following the rise and fall of your rhythms can lead to greater learning and better grades.

As a student, you have no control over the scholastic schedule. You can't attend classes only during your peak learning periods. You can't reschedule an exam falling on a critical day. But you can look to your biorhythm cycles for help.

The plus days of the rhythms, especially the intellectual rhythm, are the best days for learning. Your mind is more receptive to the stimulation of new material. Studying or review may be boring during the plus days. Try to take frequent breaks and even change rooms to provide needed stimulation.

The minus days are better left for review of learned material. Subjects such as languages that involve simple repetition or history and geography, which require memorization, are good to tackle during these days. If you are involved in learning rather than review during minus days, take careful notes in class. Use examples to illustrate important points. Don't be afraid to ask questions. Review your notes later in the day to make certain that your understanding of the material is clear. Study at your own pace. You might try studying as soon as you wake up in the morning. Get in fifteen or twenty minutes of study while your mind is fresh and alert.

Taking an exam on a critical day or during a minus period need not mean instant disaster. For a physical critical day, make sure you get plenty of sleep the night before. Don't burn the midnight oil trying to cram. You might possibly fall asleep in the middle of the exam the next day. For an emotional or intellectual critical day, the advice is be prepared. Make sure you have put in plenty of study time. Nothing will rattle you more than going into an exam without adequate preparation, especially when you are not at your best biorhythmically. If any materials are required for the exam, set the materials out the day before so you won't forget to bring anything. During the exam, read every question twice before you even start to answer. Close your eyes and take several deep breaths frequently to help fight fatigue and tension. Review the exam thoroughly before you turn it in.

Biorhythm can help make the most of your learning years. But the benefits are much more far-reaching.

In sports, biorhythm can be your own personal coach. The plus days of the physical rhythm are the best days for intensive athletic training. You'll have the energy and stamina to endure long workout sessions. Start to wind down your training as your physical rhythm inches toward the minus days. Avoid overtraining during the minus phase and conserve your strength for the next upswing into the plus days. Beware of the instability of critical days, which may lead to accidents or injuries during training or sporting events.

The same advice holds true for the varied physical leisure activities that are so popular today. From bicycling to camping and mountain climbing, it is most important to be aware of your biorhythmic conditions, whether plus, minus, or critical. Learn to be happier, healthier, and safer.

Safety on the job is an important consideration for industry as well as for the worker. Accidents have been shown to occur most often on critical days. In companies that have adopted biorhythm programs, the accident and injury rate for workers has dropped considerably. Workers in companies without biorhythm programs should try to get such programs started. Even those with part-time jobs such as baby-sitting, bagging groceries, or pumping gas can still take advantage of biorhythm's benefits by calculating their own cycles to be more aware of accident-prone critical days.

Whether we are on the job or in school, on the sports field or at leisure, biorhythm can help in our relationships with those around us. Compatible persons have harmonious relationships. They experience the ups and downs of the three rhythms at the same time. Friends who are biorhythmically incompatible can use biorhythm as a base for understanding that the harsh word or the emotional outburst was the result of conflicting rhythms. In that way, the relationship remains intact while the problem of conflicting rhythms can be handled with kindness and consideration.

The benefits of biorhythm are far-reaching, not just in the present, but also in the future. The world of tomorrow may show us all living according to the biorhythm chart. Doctors will choose the best day—for themselves and their patient—on which to operate. Education will stress learning during the peak intellectual period. Workers will be reassigned jobs on critical days to avoid accidents. Parents may be able to determine the birth date and sex of their child even before conception. These marvelous possibilities are in the future. But the only time that really counts is right now.

Take advantage of your positive days; compensate for your weak times. Learn to understand yourself and those around you. Begin to know the person you really are. This knowledge can give you the power to take charge of your life. All this is possible by discovering your natural ups and downs through biorhythm.

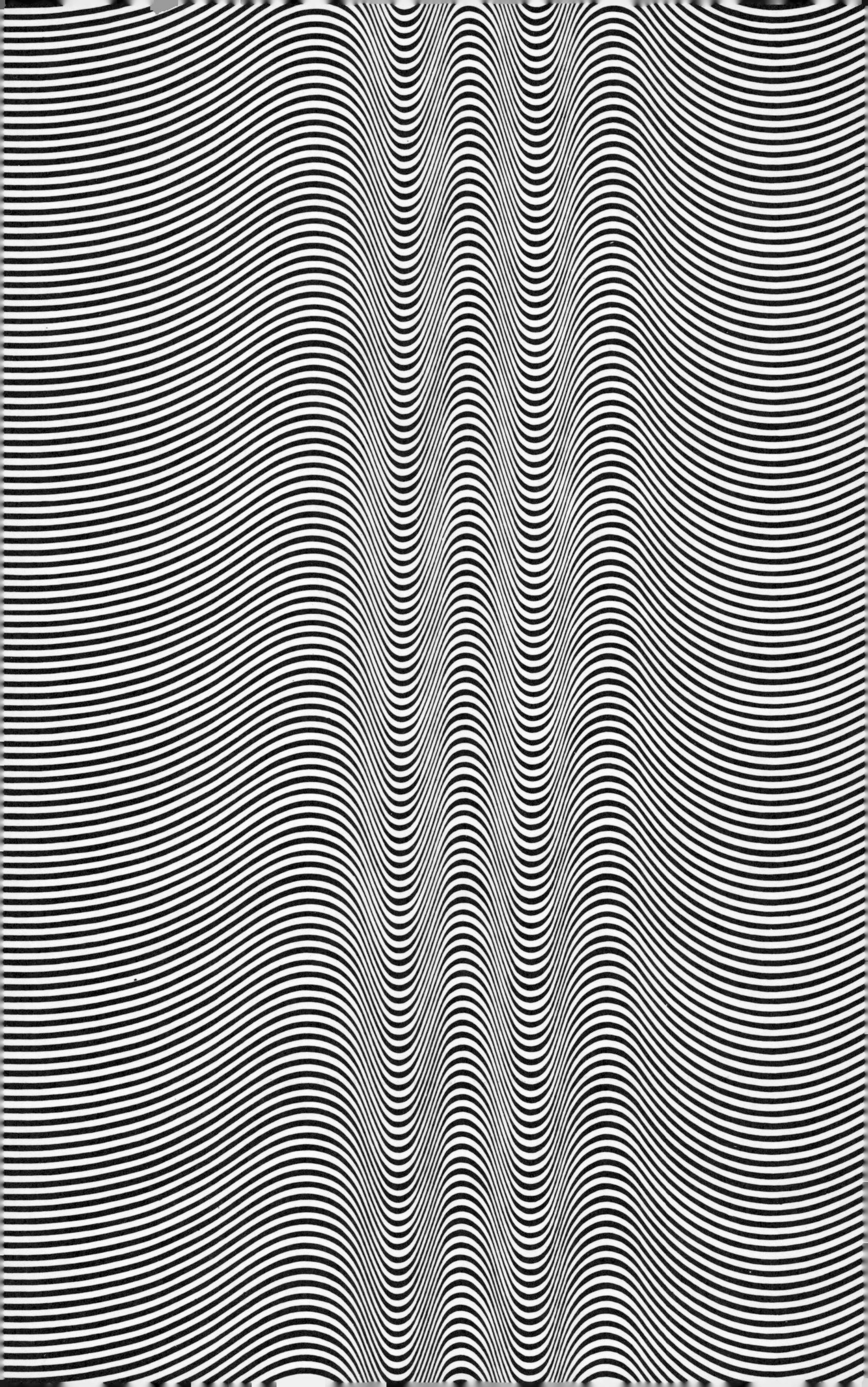

BIORHYTHM OBITUARIES OF FAMOUS PEOPLE

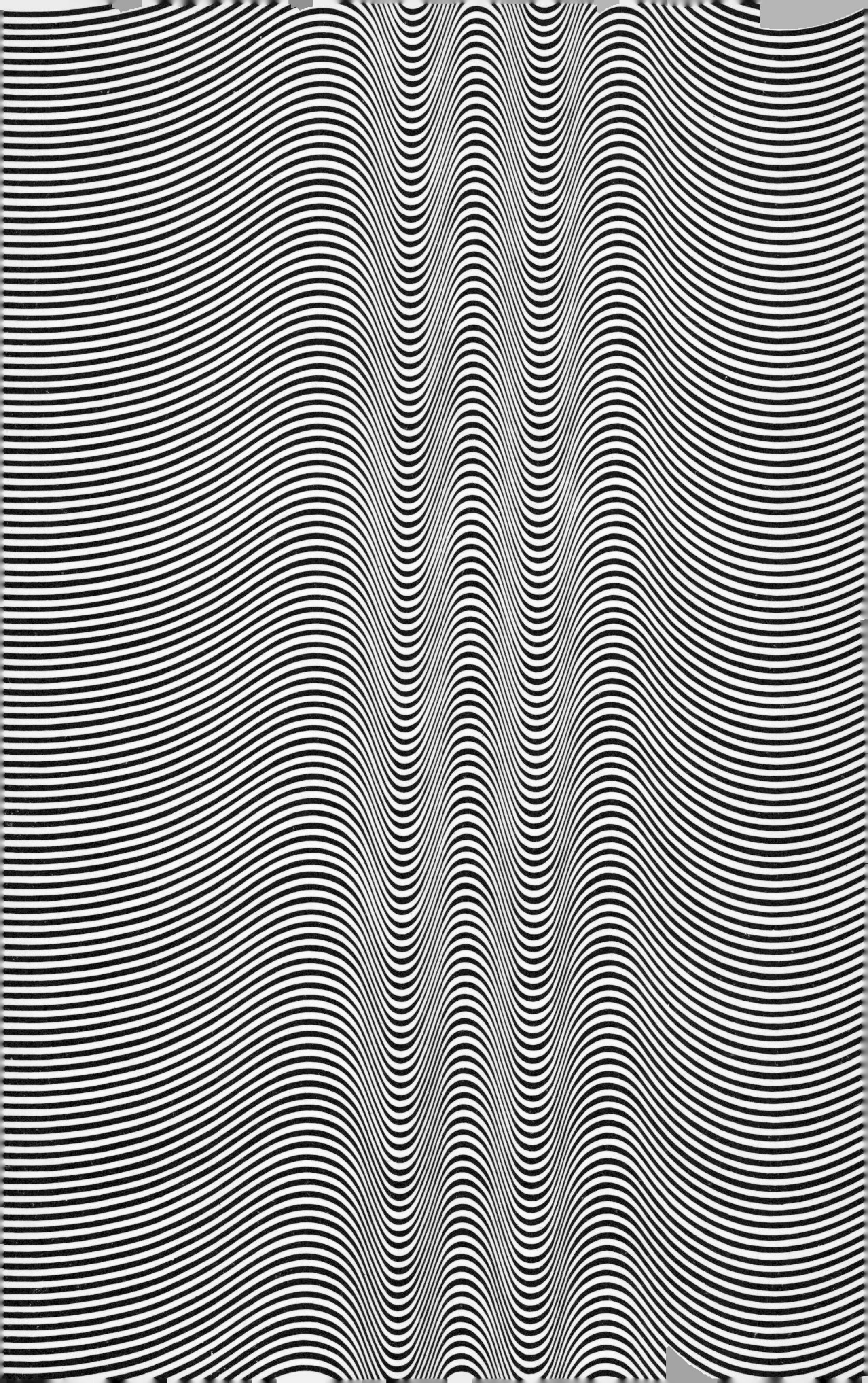

Clark Gable, born February 1, 1901, was best known for his portrayal of Rhett Butler in the movie *Gone With the Wind.* He suffered a heart attack on November 6, 1960, after experiencing a physical critical day. He died on his next critical day, November 16, 1960.

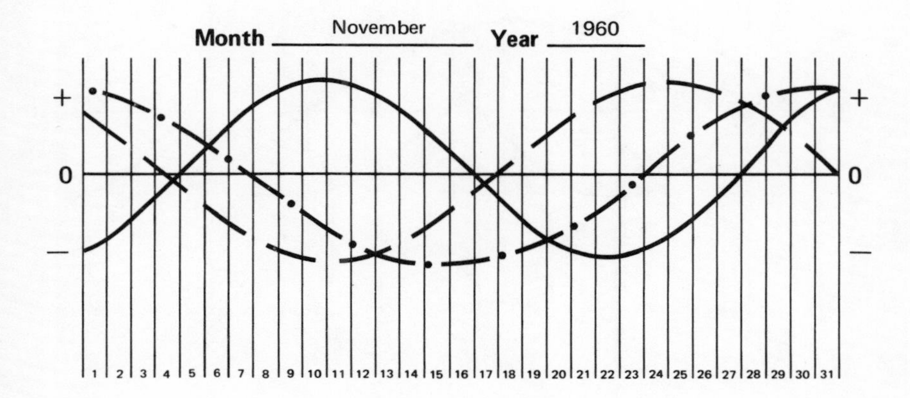

Freddie Prinze shot himself on January 28, 1977, when both his physical and emotional rhythms were in the minus phase. He died the following day, which was a critical day in the physical rhythm. The comedian-actor, born on June 22, 1954, was the co-star of the television series *Chico and the Man.*

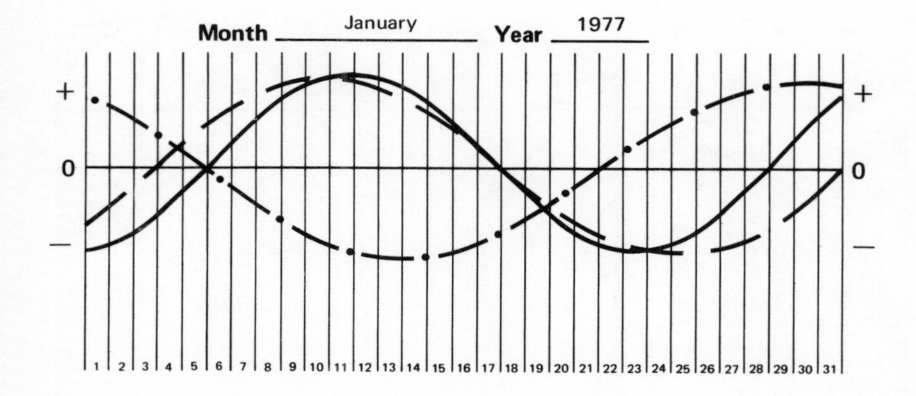

Haile Selassie I, Emperor of Ethiopia, was born on July 23, 1892. He died, on August 27, 1975, one day after a physical critical day.

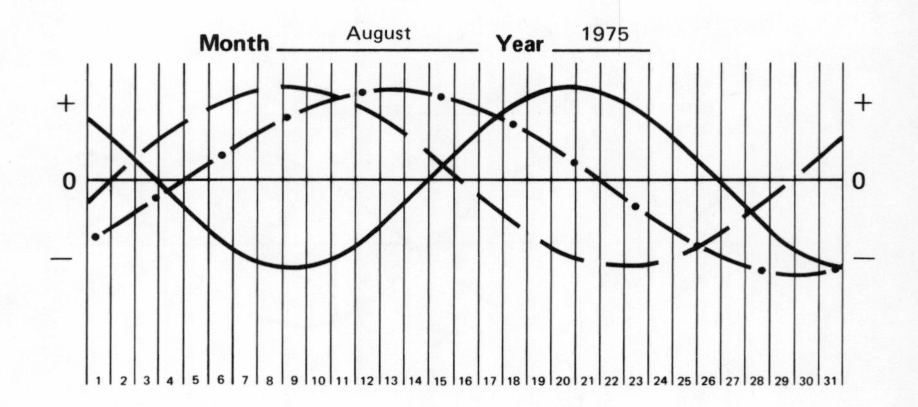

Sculptor, artist, and inventor of the mobile, Alexander Calder was born on July 22, 1898. His death on November 11, 1976, followed a physical critical day.

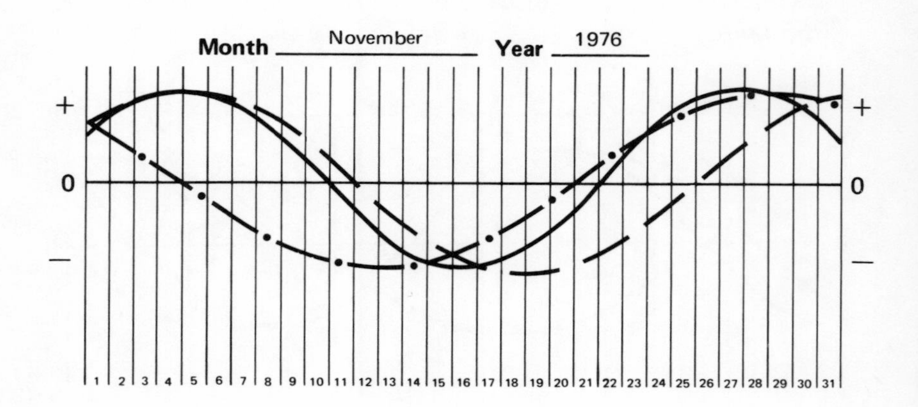

Chet Huntley, born on December 10, 1911, was a television journalist for NBC. He died on March 20, 1974, right before a physical critical day.

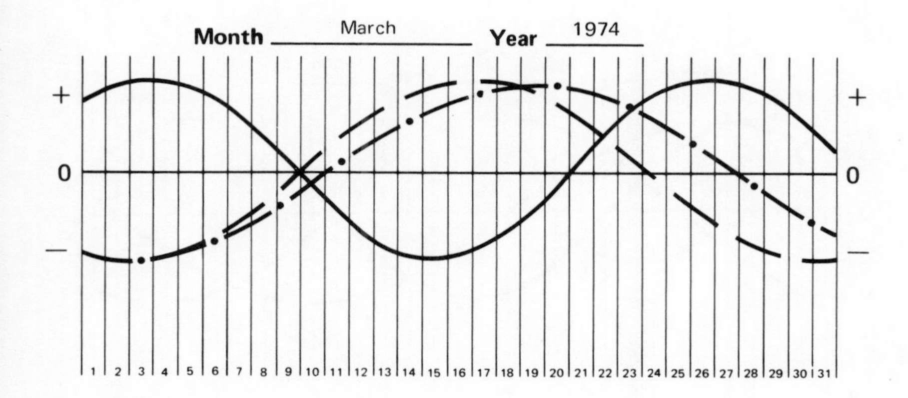

Rod Serling was born on December 25, 1924. He was a television writer, producer, and narrator of the television series *Twilight Zone.* He died on June 28, 1975, on an intellectual critical day. A physical critical day occurred on June 27, and his emotional rhythm was low.

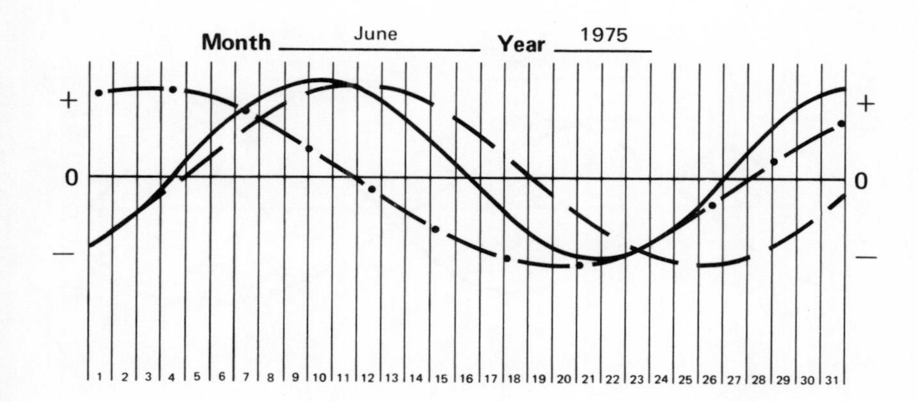

James Jones, author of *From Here to Eternity,* was born on November 6, 1921. He died of a heart attack on May 9, 1977. His chart shows an emotional critical day on May 8 and a physical critical day on May 10.

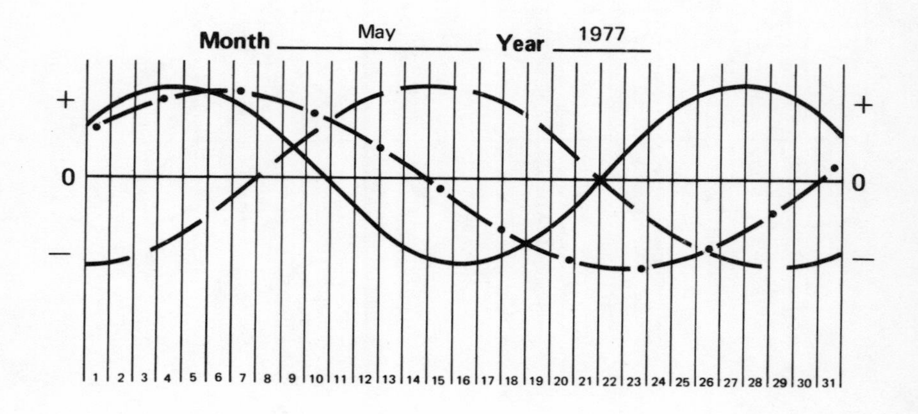

Author Jacqueline Susann was born on August 20, 1921. She died on an emotional critical day—September 21, 1974. She was also under the influence of a physical critical day, which occurred on the twentieth.

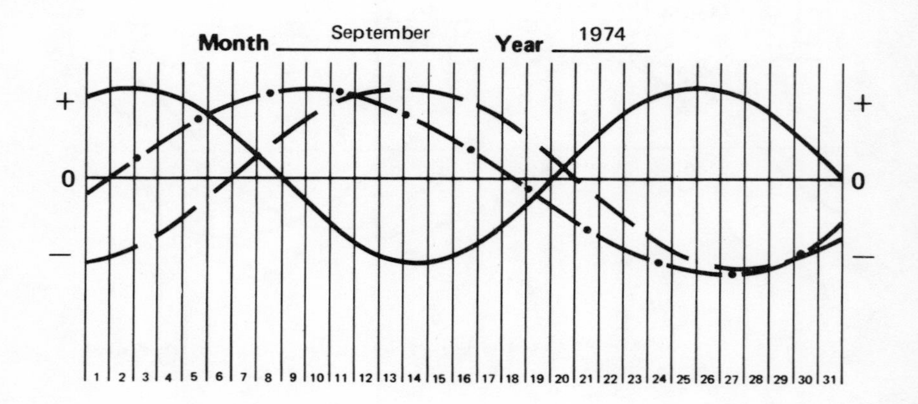

British author Agatha Christie was born on September 15, 1890. Her death on January 12, 1976, was on an emotional critical day. Her physical rhythm was at the last day of its cycle.

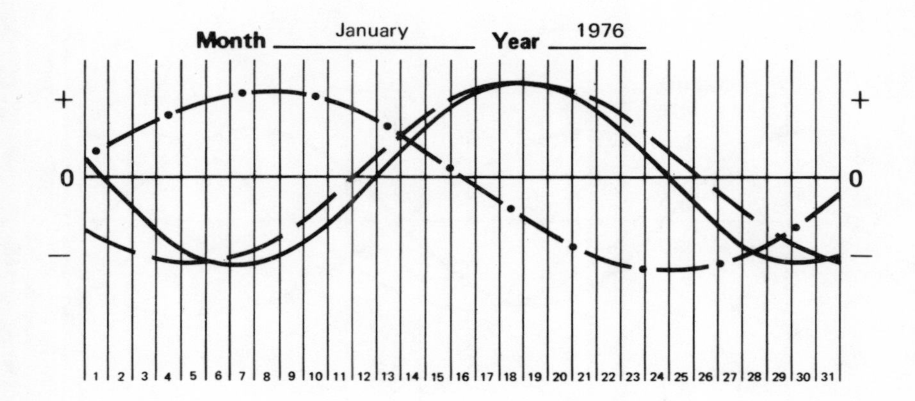

Actor Peter Finch, born on September 28, 1916, received a posthumous Academy Award for his role in the movie *Network*. He died of a heart attack on January 14, 1977, just one day after a double critical day.

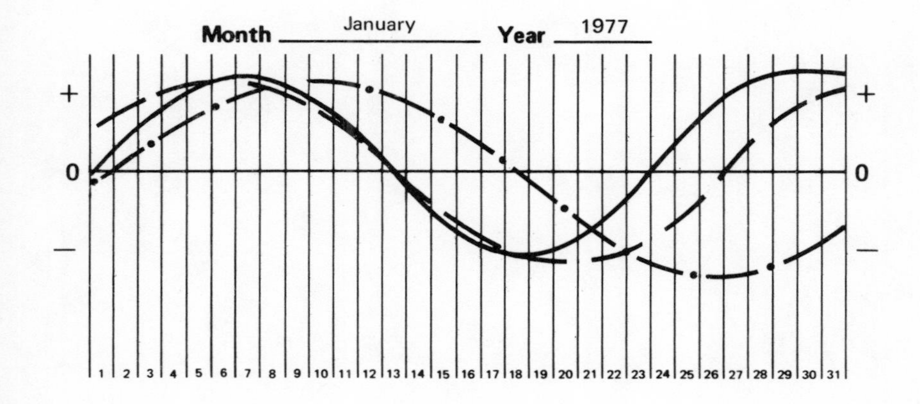

Richard Tucker, leading tenor with the Metropolitan Opera, died of a heart attack on January 8, 1975. He was born on August 28, 1913. His death occurred on a physical critical day and one day before a critical day in the emotional rhythm.

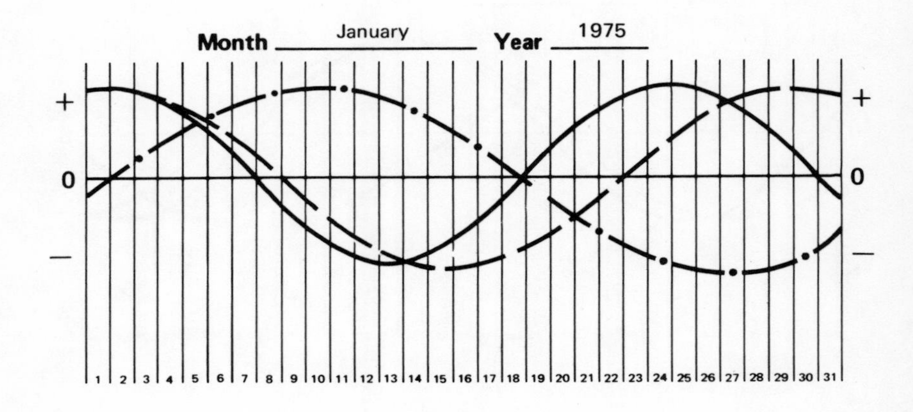

CHART OF YEARS

(Leap years are indicated by an asterisk)

1900	1923	1946	1969	1992*
1901	1924*	1947	1970	1993
1902	1925	1948*	1971	1994
1903	1926	1949	1972*	1995
1904*	1927	1950	1973	1996*
1905	1928*	1951	1974	1997
1906	1929	1952*	1975	1998
1907	1930	1953	1976*	1999
1908*	1931	1954	1977	2000*
1909	1932*	1955	1978	2001
1910	1933	1956*	1979	2002
1911	1934	1957	1980*	2003
1912*	1935	1958	1981	2004*
1913	1936*	1959	1982	2005
1914	1937	1960*	1983	2006
1915	1938	1961	1984*	2007
1916*	1939	1962	1985	2008*
1917	1940*	1963	1986	2009
1918	1941	1964*	1987	2010
1919	1942	1965	1988*	2011
1920*	1943	1966	1989	2012*
1921	1944*	1967	1990	2013
1922	1945	1968*	1991	2014

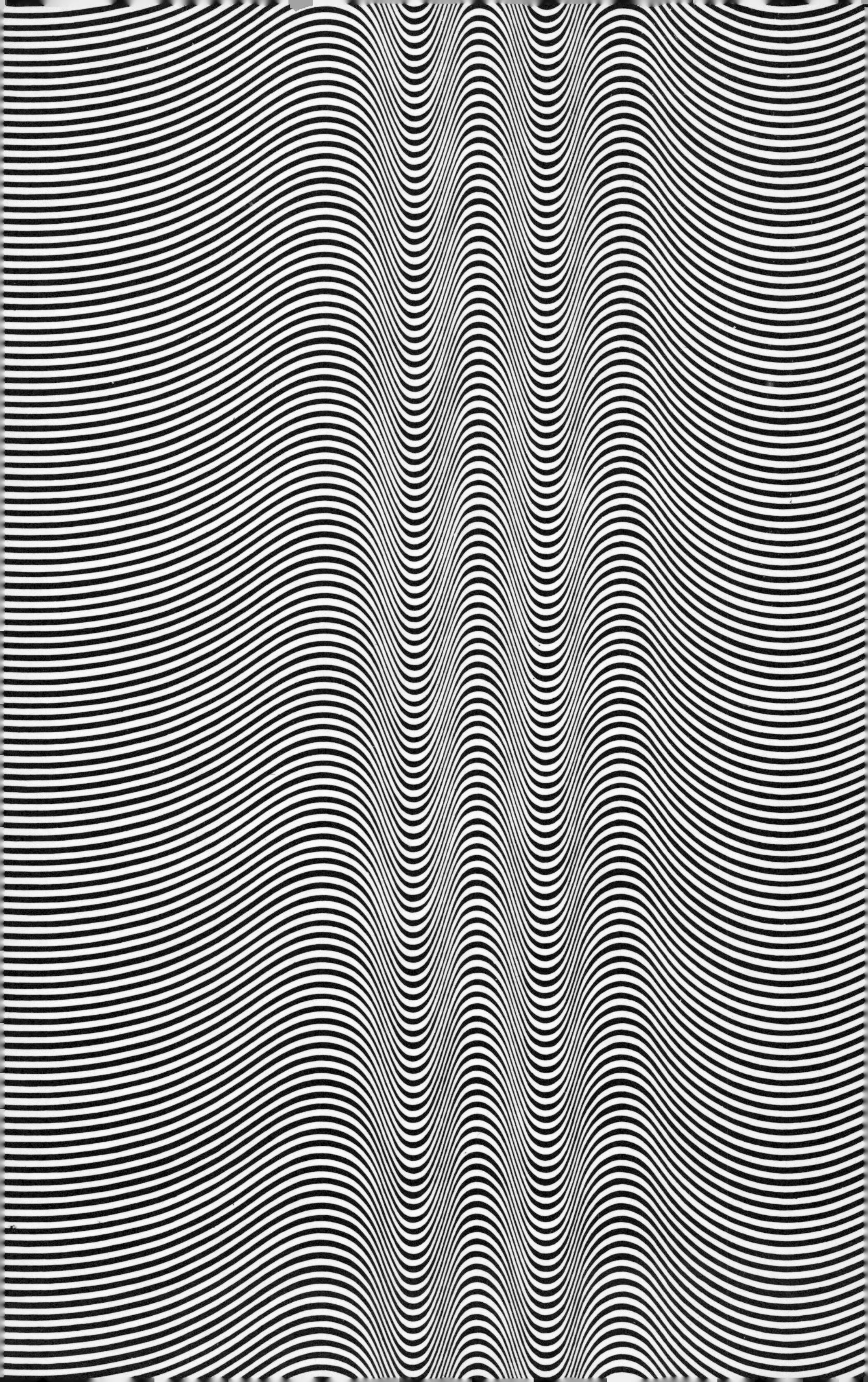

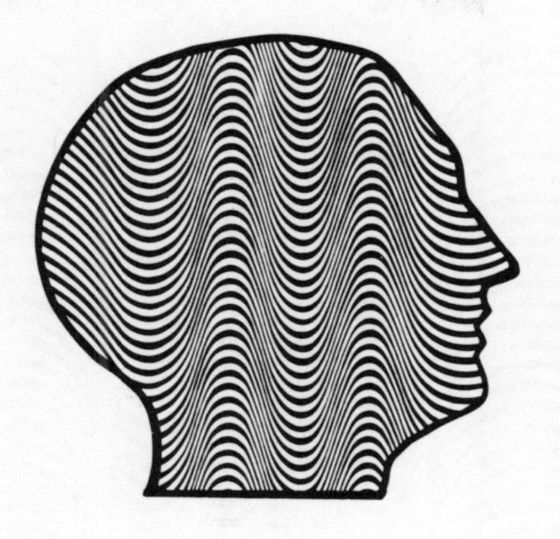

BIBLIOGRAPHY

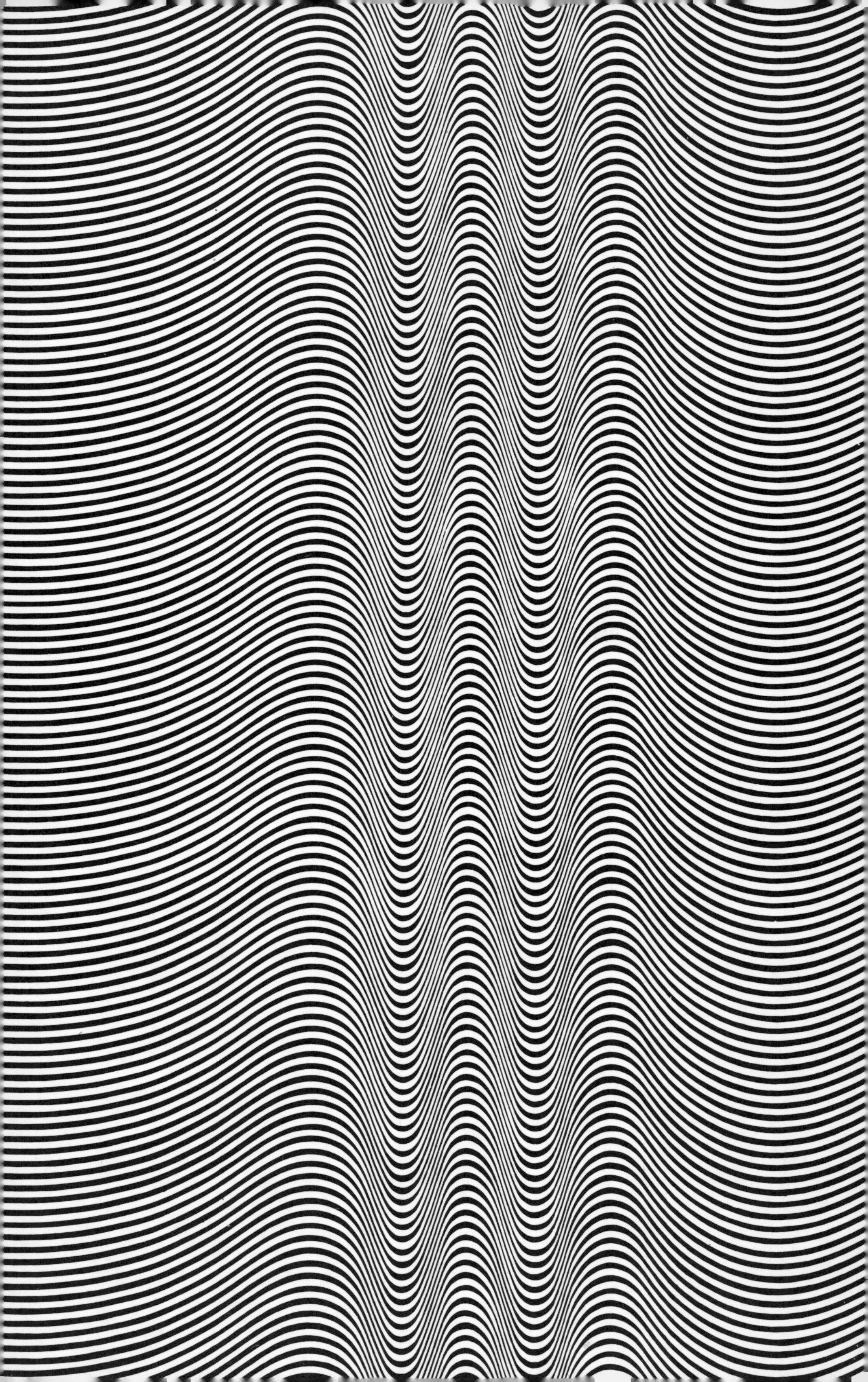

BOOKS

Advances in Experimental Medicine and Biology, Volume 54—Biological Rhythms and Endocrine Function. New York and London: Plenum Press, 1975.

Bunning, Erwin. *The Physiological Clock.* 3d ed. London: The English Universities Press Ltd; New York, Heidelberg, Berlin: Springer-Verlag, rev. 1973.

Conroy, R. T. W. L., and Mills, J. N. *Human Circadian Rhythms.* London: J. and A. Churchill, 1970.

Gittelson, Bernard. *Biorhythm: A Personal Science.* New York: Warner Books, 1975.

Luce, Gay Gaer. *Body Time.* New York: Pantheon Books, 1971.

Luce, Gay Gaer, and Segal, Julius. *Insomnia: The Guide for Troubled Sleepers.* Garden City, N.Y.: Doubleday and Company, 1969.

Pengelley, Eric T. ed. *Circannual Clocks: Annual Biological Rhythms.* New York, San Francisco, London: Academic Press, Inc., 1974.

Reinberg, Alain, and Ghata, Jean. *Biological Rhythms.* New York: Walker and Company. First pub. France, 1957; trans. including new material, 1964.

Smith, Dr. Robert E. *The Complete Book of Biorhythm Life Cycles.* New York: Aardvark Publishers, 1976.

Strughold, Hubertus. *Your Body Clock: Its Significance for the Jet Traveler.* New York: Charles Scribner's Sons, 1971.

Tatai, Kichinosuke. *Biorhythm for Health Design.* Japan Publications, 1977.

Thommen, George S. *Is This Your Day?* rev. ed. New York: Crown Publishers, 1973.

Ward, Ritchie R. *The Living Clocks.* New York: Alfred A. Knopf, 1971.

Wernli, Hans. *Biorhythm: A Scientific Exploration into the Life Cycles of the Individual.* New York: Crown Publishers, 1961.

Wood, Clive. *Human Fertility: Threat and Promise.* New York: Funk and Wagnalls, 1969.

ARTICLES

Arehart-Treichel, Joan. "Attention, Biological Clockwatchers." *Science News,* Sept. 7, 1974, pp. 156–157.

"Biorhythms in the Sky." *Science Digest,* June 1975, p. 16.

Fischer, Dr. Irving C. "Can You Choose The Sex of Your Child?" as told to Joseph Kaye, *Ladies' Home Journal,* Feb. 1962.

" 'In Spring a Young Man's Fancy' . . . Doctors Say It's All a Myth." *U.S. News and World Report,* May 17, 1976, pp. 64–67.

Lavie, Peretz, and Kripke, Daniel F. "Internal Tempos II Ultradian Rhythms: The Ninety Minute Clock Inside Us." *Psychology Today,* April 1975, pp. 54, 55, 65.

Luce, Gay Gaer. "Trust Your Body Rhythms." *Psychology Today,* April 1975, pp. 52–53.

"Switched-On Membranes: Internal Clocks." *Science News,* July 10, 1976, p. 23.

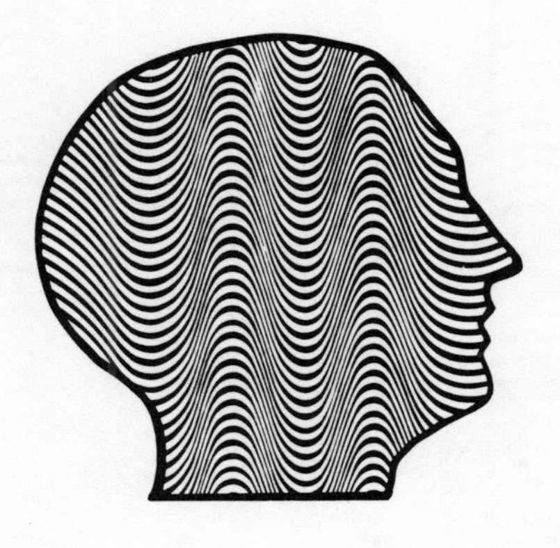

INDEX

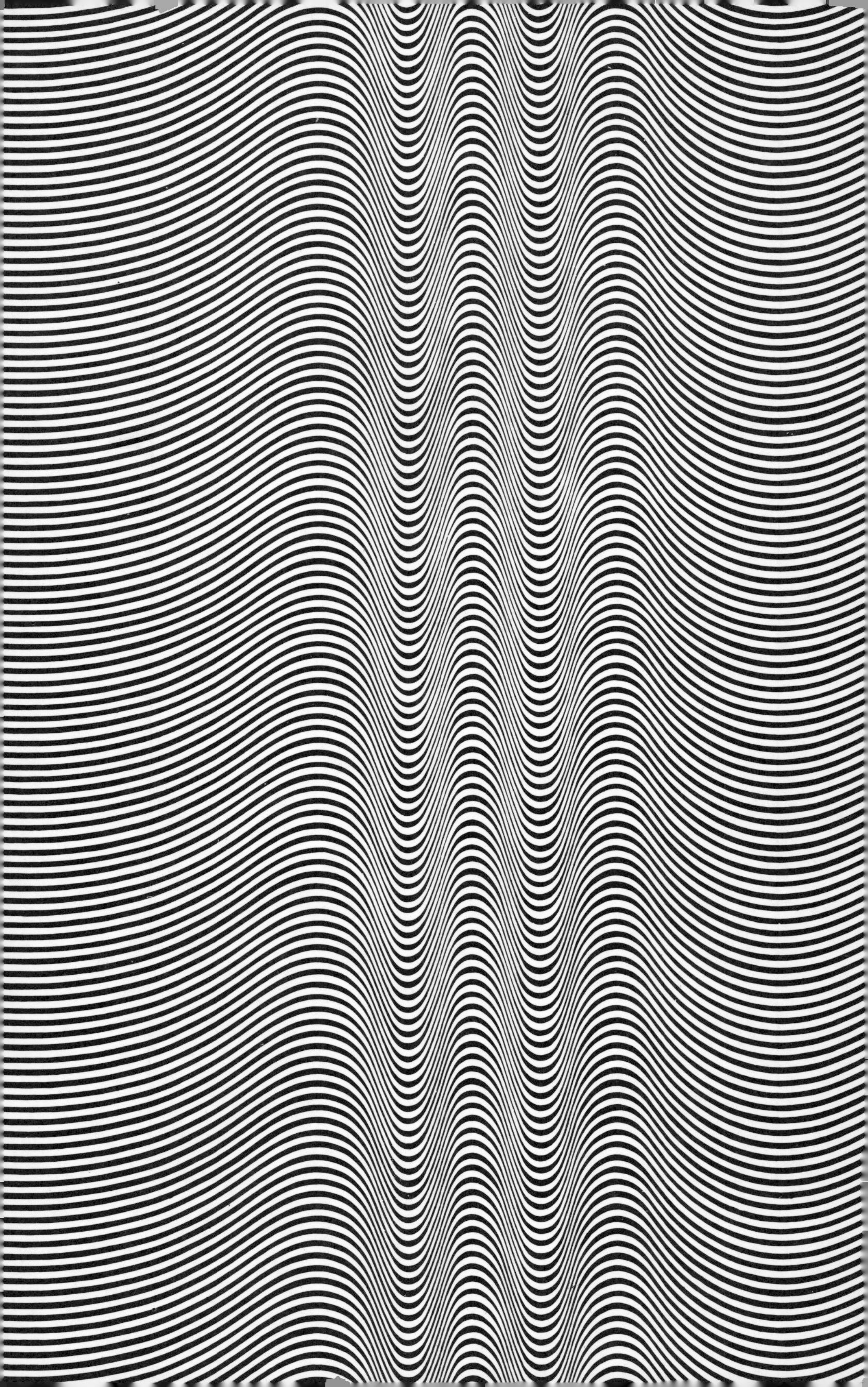

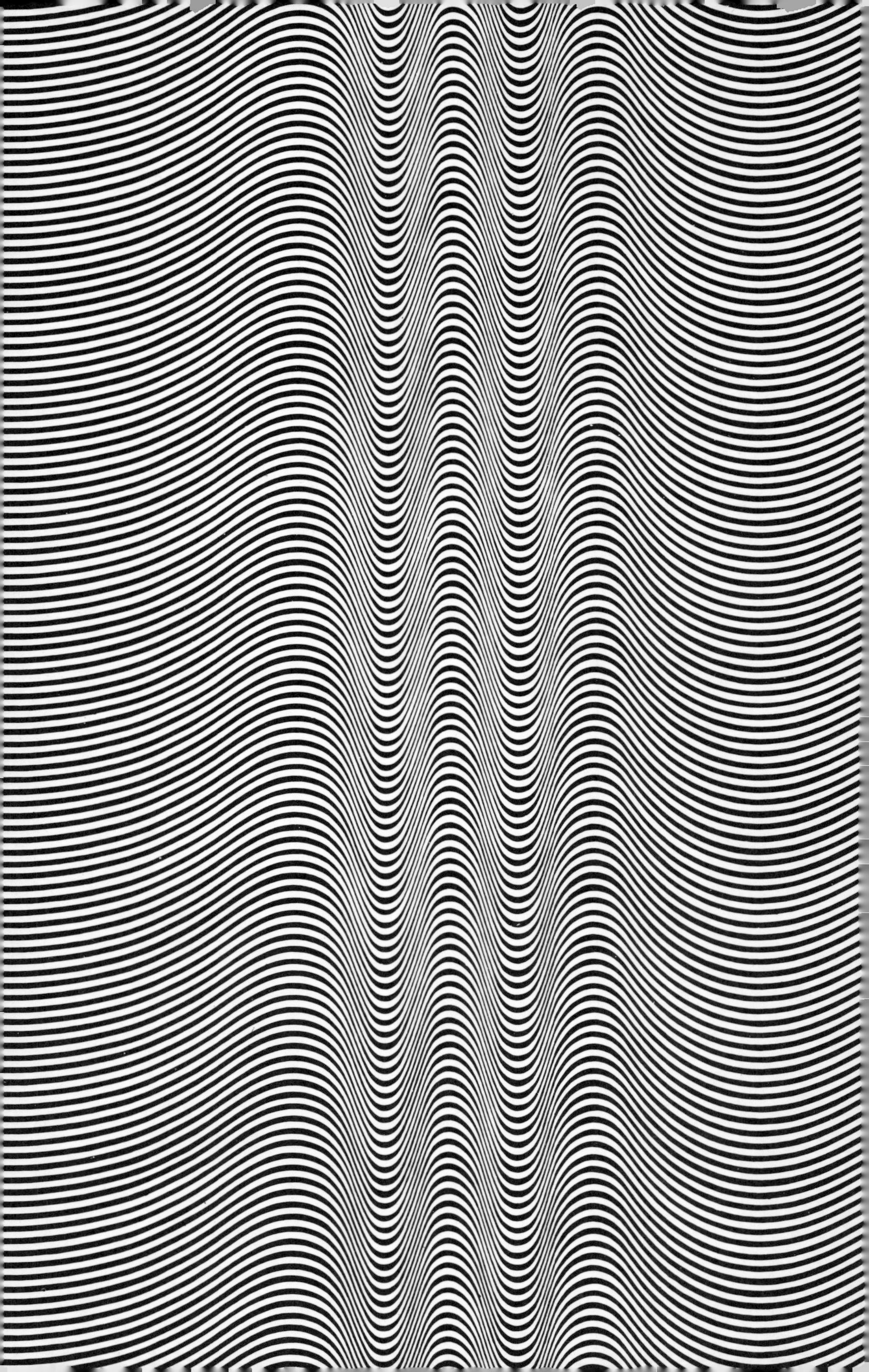

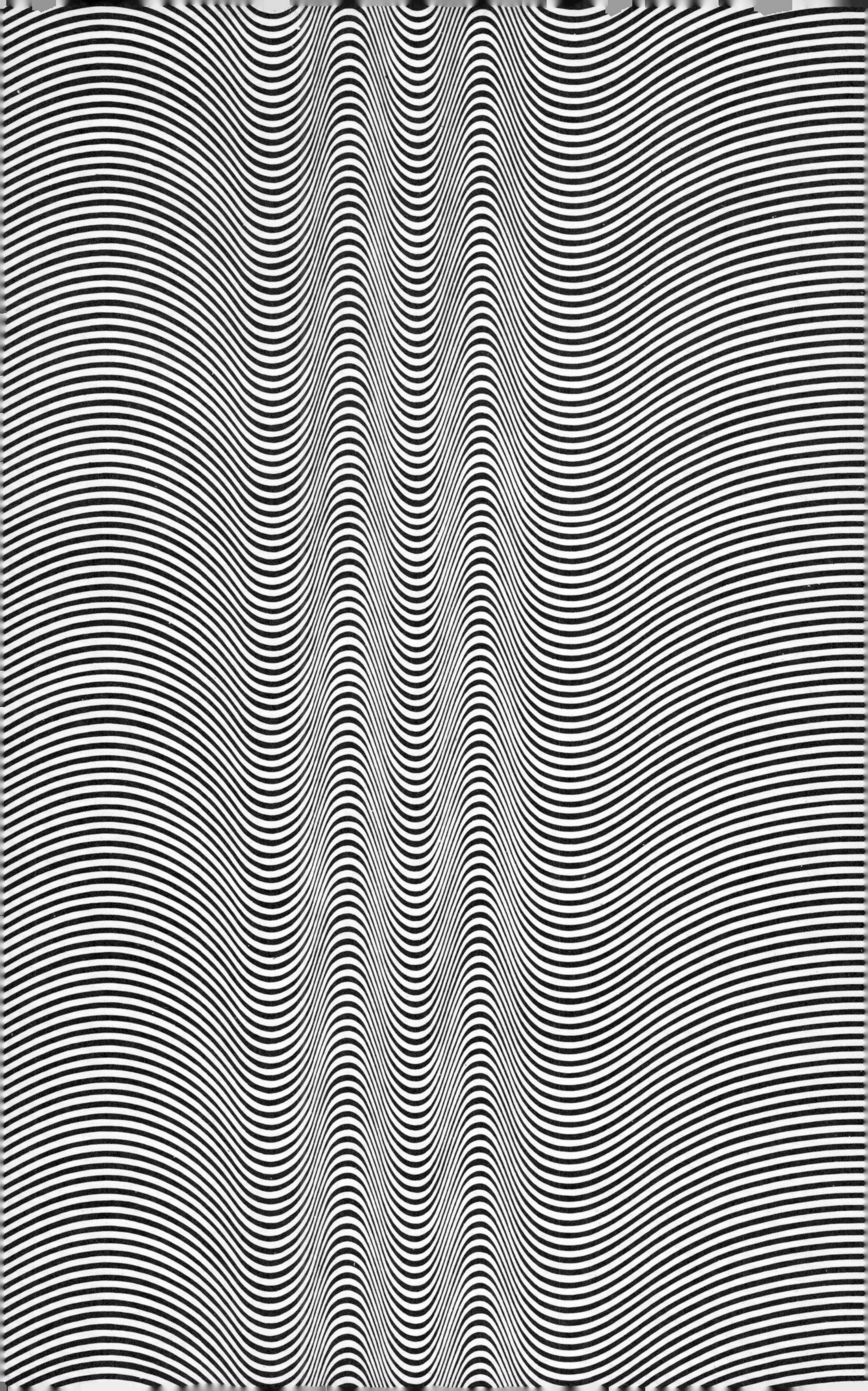

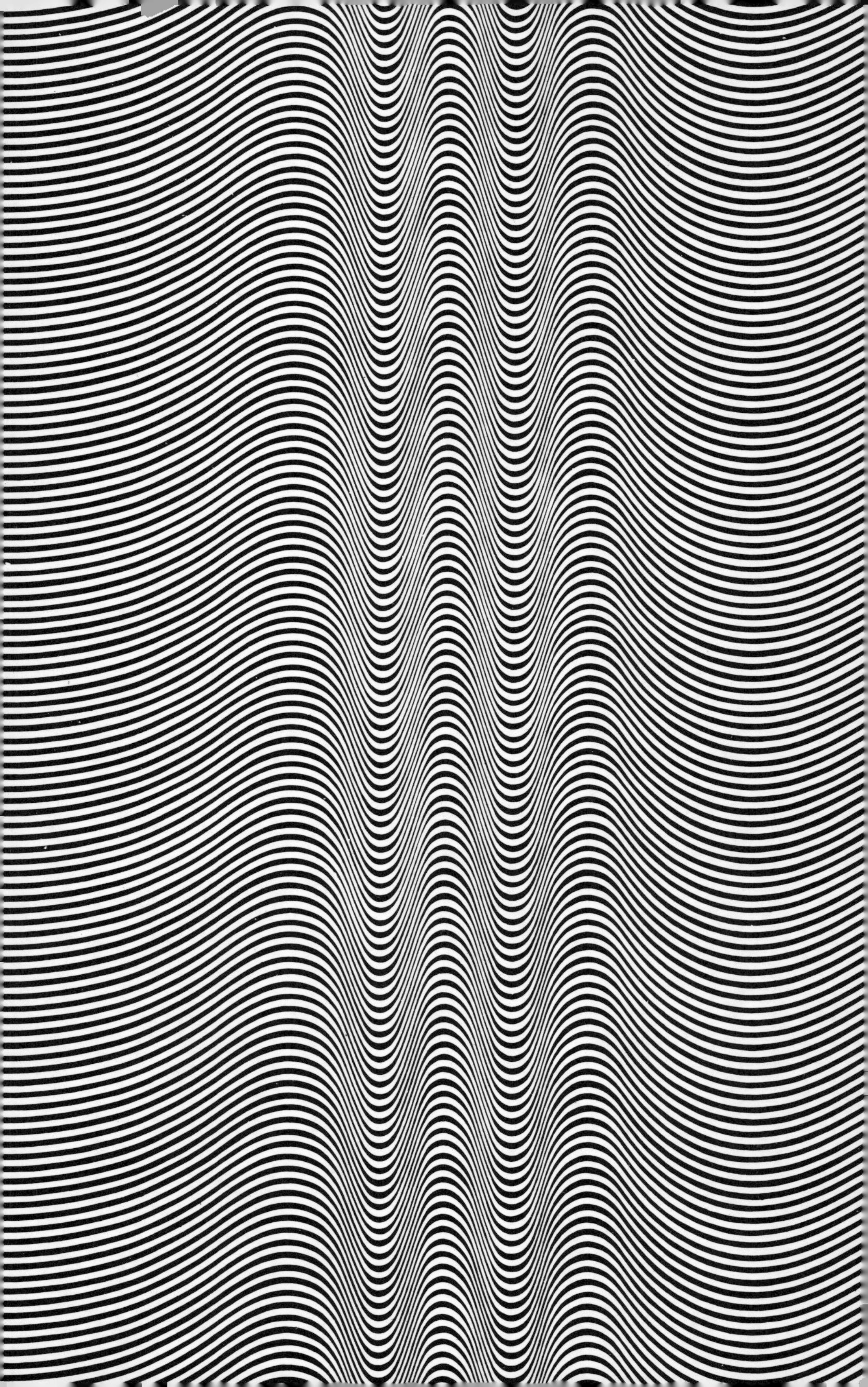

ABOUT THE AUTHOR

Pauline Bartel applied biorhythm theory to the writing of this book. She wrote during the plus phases of her physical, emotional and intellectual rhythms, and did research during the minus phases. The author lives in Albany, New York, with her husband, and works as a medical secretary. This is her first book.

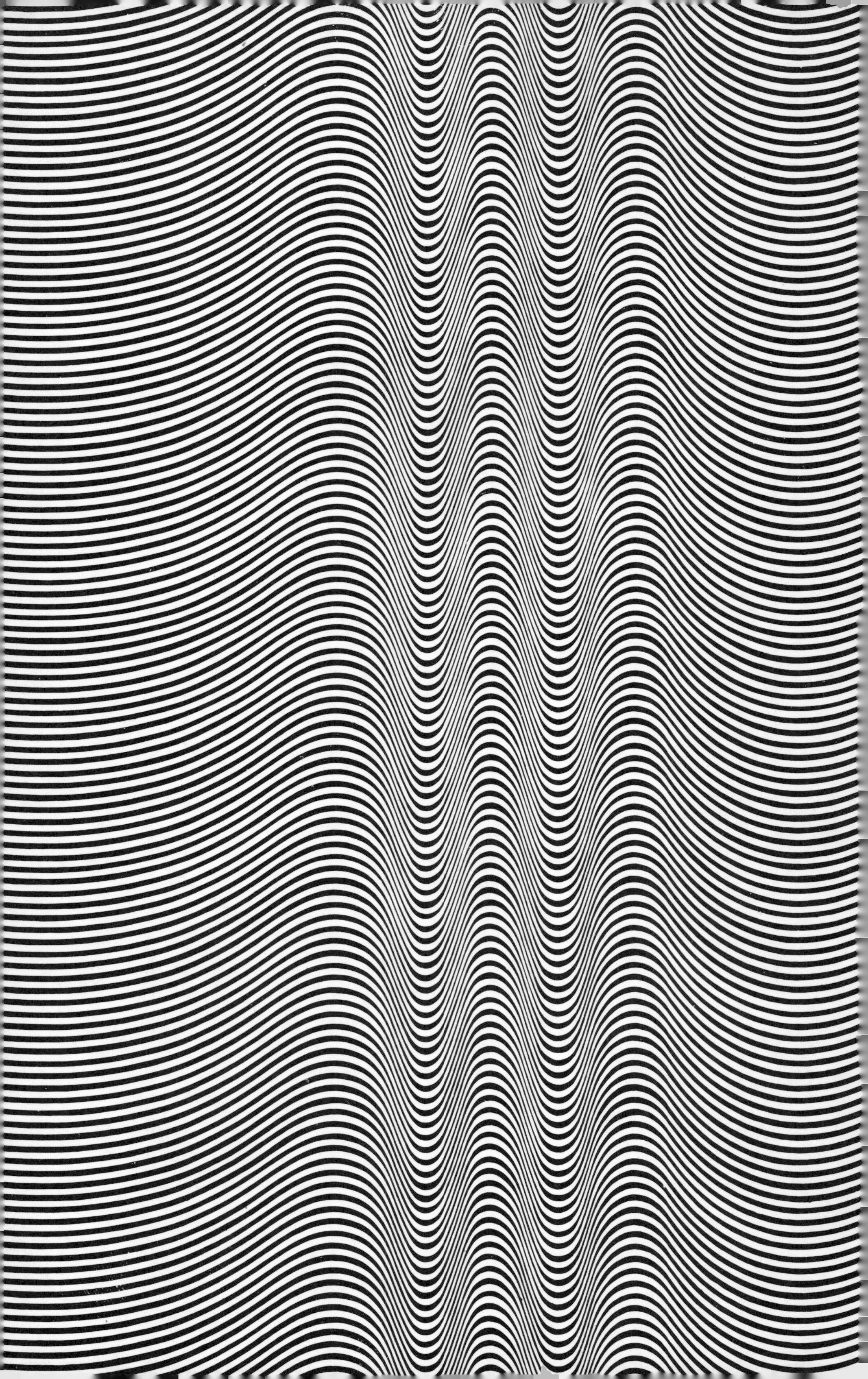